MAKING TV PAY OFF

A Retailer's Guide to Television Advertising

by HOWARD P. ABRAHAMS

FAIRCHILD PUBLICATIONS, Inc.
New York

Standard Book Number: 87005-140-7

Printed in the United States of America

TABLE OF CONTENTS

DECISIONS TO MAKE

OTHER TELEVISION GUIDE LINES

INTRODUCTION

There is a temptation for an author to have his work produced as a typographical beauty on slick stock and magnificently bound. Such a book would be skimmed through, take a place of honor on the library shelf and possibly removed from time to time for points of reference.

Not this book.

Having worked with hundreds of retailers and other local advertisers, the author knew their problems in using TV -- the sophisticated advertiser as well as the neophyte. He found they all needed a step-by-step plan to guide them through a series of decisions. This workbook is the result. This workbook takes the advertiser from decisions on the goals to have on television, how to achieve these goals with decisions on budgeting, scheduling, working with vendors, creating the commercial, buying the time and, finally, checking the results.

The decisions are made easy through a series of check lists. So sharpen up a bunch of pencils. Then start reading.

Howard P. Abrahams

TODAY'S CUSTOMER IS A DIFFERENT CUSTOMER . . .

TO REACH TODAY'S CUSTOMER REQUIRES A DIFFERENT METHOD

At one time in the past, the central city was the heart of all business activity. This included:

- department stores
- specialty fashion stores
- variety stores
- furniture stores
- men's stores
- automotive dealers
- other retail operations

The central city was the heart of:

- theaters
- sports events, museums
- all forms of entertainment

The central city housed the leading business activities:

- banks and other financial institutions
- business offices
- factories and other working places
- hotels

The masses of population were tied directly to the central city. Along with those who lived on the fringes of the central city were the people who came "downtown" to work, to shop, to conduct their business and to seek entertainment.

The customer was different due to:

- larger families
- loyalties to specific stores for specific merchandise
- shopping centered on necessities of life rather than luxuries
- greater extremes in income -- rich and poor

The old customer was influenced into buying by family tradition and word of mouth. Advertising was done in the city newspapers and in some direct mail. With the shopping areas concentrated in the central city, good locations were of prime importance to create traffic. Window displays played a big part in motivating people from the sidewalks into the stores.

Changes in living locations and business locations came about gradually. Since World War II, population shifts, meaning customer shifts, created new customer habits and new customer patterns:

- non-stop movement of customers from city to suburbs
- many suburbs became their own "central city" with their own suburbs and with people moving further out into the exurbs
- rural population moving into the metro areas

The customers who made the shift from city to suburbs took on different characteristics from the city dwellers due to:

- younger families
- higher education levels
- higher income levels
- greater ownership of goods -- houses, autos, home furnishings, appliances, clothing, everything

The new customer, along with the city customer, developed new buying habits:

- greater use of credit -- both store credit and bank credit. Credit sales exceed cash sales in most big stores.
- lack of loyalty to specific stores because of sameness of merchandise and merchandise prices, store appearance, and lack of service.
- easy to get to stores because of car mobility on improved highway systems. This made more stores and shopping centers available.

Businesses of all kinds responded to the population movement and the new customer:

- with branch operations, they moved where the customer was situated.
- continued to maintain their downtown "flagship" stores, often with larger selections in certain departments. They appealed to the city dweller, but also shoppers from the suburbs who enjoyed downtown shopping trips.

Many suburban customers rarely go downtown to shop, conduct business, or to be entertained. Their interest in the downtown newspaper is also decreasing. To reach these new customers, there are several choices.

- suburban newspapers. Inexpensive per copy, but they add up considerably when several are needed to insure coverage. Cost per thousand generally runs high compared to large city papers. Additionally, there is the problem of print production for multi-papers.

- direct mail to store lists, bought lists, and occupancy lists. Production costs and constantly higher mail costs bring about problems.
- radio. This medium can be excellent, especially when a store is to reach the teens and the younger age groups. Preparation of radio advertising is relatively easy and inexpensive. Fractionalization of audience, due to the large number of stations per market, is a drawback.
- television. Here, both the city and the suburban customer can be reached at the same time with one message.

Advertising was once defined as "salesmanship in print." This is certainly true of newspapers.

Television is better defined as a personal selling medium. TV has specific advantages.

- almost everyone has a television and views it many hours a day. Household viewing averages over six and one-quarter hours a day, an all-time high. Your local station can give you up-to-date figures.
- combines the selling advantages of sight, sound, motion, demonstration, and color.
- involves emotion to move people to buy or accept certain characteristics about the advertiser.
- reaches hugh masses of customers in a wide radius -- city, suburbs, exurbs, all at the same time.

With television you select the people you want to reach and when to reach them:

- with TV you select the people you require for your company, or for certain departments, for specific items or certain services and other features
- different people view at different times of the day
- different people watch different shows
- you select the age, sex and other demographic qualities of the customers you want and match the shows and time of day to the time when these customers are available and when they view

Your opportunity to reach customers on television is greater because of the amount of time they devote to the medium.

	Minutes Per Day Watching Television	Minutes Per Day With Newspapers
Young Women	155	26
High Income Women	131	38
Young Men	124	28
High Income Men	101	36

WHY STORES AND OTHER LOCAL BUSINESSES USE TELEVISION

	Check the Important Reasons for Your Store
Reach both city and suburban customers at the same time.	______
Reach suburban people who only read city newspapers.	______
Reach young customers who watch more and read less.	______
Project the store image, or create a brand new image. Examples: To sell the idea of the store to young people. Many banks use TV to give a feeling of friendly service in contrast to the big, unfriendly institution.	______
Overcome "bigness" of a large store or enable a small store to give the effect of a larger, progressive retailer. Personalities on TV can get either impression across.	______
Promote services and other institutional services. Effectiveness is difficult in print. Television can sell credit, easy parking, large assortments, liberal exchanges, friendly salespeople.	______
Introduce new departments and services.	______
Create traffic through special value and timely items.	______
Create traffic by promoting departmental or store-wide events.	______
Promote items needing demonstration. Examples: Sewing machines, power tools, paints, reupholstering, remodelling, and home improvement.	______

Check the Important Reasons for Your Store

<u>Flexibility of merchandise</u>. Examples: Weather turns cold. Switch from the scheduled commercials to cold weather merchandise. Weather turns warm. Switch to fur storage, swim suits, outdoor sports. Many stores keep "stand-by" commercials for such opportunistic situations. _____

<u>Impress the financial community</u>, or vendors, or salespeople. Television, a modern medium associated with important companies, shows their people that the store is modern. _____

<u>Increase the media mix</u> to make all media work with each other for greater impact on the customer. _____

Local Advertisers Along With Manufacturers

Have Constantly Increased Their TV Advertising

	Total	Network	Spot	Local
1950	171	85	31	31
1960	1,627	820	527	280
1970	3,596	1,658	1,234	704
1978	7,550	3,000	2,270	2,080

1978 is an estimate. Amounts in millions of dollars.
Add (000,000)

So you have decided to use television to build sales . . .

Big questions originate:

- What to use television for
- How to produce commercials
- How to budget
- How to buy the right schedules
- Many other questions

Specific decisions must be made.

DECISION #1

Determine Your Company's Goals

The successful company has fixed goals, both short and long range. When these goals are accepted by all executives, a definite direction is fixed for the entire company to follow.

Of course, goals change from time to time depending upon any number of circumstances, such as when opportunities for expansion originate, when the economics of the nation change, when the immediate location of the stores change, or when new developments in retailing take place, including new competition.

Along with a determination to use television, the store's goals should be examined and discussed by top management in the store. When the goals are crystal clear in the minds of the management team, it is then possible to take the next step.

DECISION #2

Determine Your Television Goals

It is a rare situation when all the goals of the store can be accomplished through one medium of promotion, but this is particularly true of television. Decision #2 requires limiting the number of specific goals to be used on television. They may be the most important ones for the store. They may be needed to build certain departments or areas of merchandise. They may be television goals for the period immediately ahead, with a plan to shift television goals to other areas subsequently. As an aid to Decisions #1 and #2, here is a check list:

<u>Typical Goals Which Stores Use on Television</u>	✓ <u>Which of These Are Your TV Goals?</u>
Portraying the store image the way it is.	______
Our store image stands for	
________________________	______
________________________	______
Changing the image of the store.	______
Improving women's fashion image in departments of	
________________________	______
________________________	______
________________________	______
Improving men's wear image in departments of	
________________________	______
________________________	______
________________________	______
Improving image in home furnishings, such as	
________________________	______
________________________	______
________________________	______
Creating a "young feeling" especially in	
________________________	______
________________________	______
________________________	______
Other image areas	
________________________	______
________________________	______

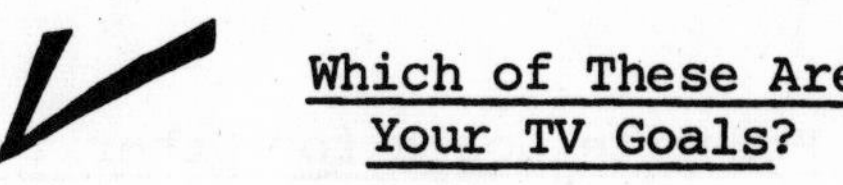

Which of These Are Your TV Goals?

Typical Goals Which Stores Use on Television

Selling service features of the store such as:

Large selections ______

Easy parking ______

Liberal returns ______

________________ ______

________________ ______

________________ ______

Add any other goals, both long range and short range, appropriate to your store.

____

____

____

____

____

____

____

____

____

____

____

Possible Goals for Other Types of Local Business

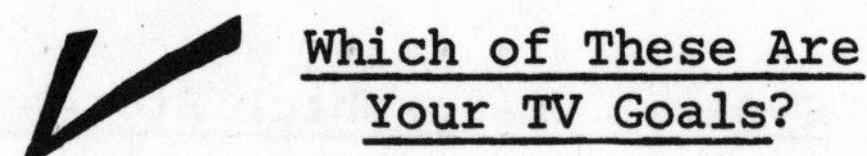

BANKS

To sell specific service features. ____

Checking accounts ____

Savings accounts ____

Christmas clubs ____

Safe deposit boxes ____

Financial advisory services ____

Mortgage and other loans ____

To sell a broad image of the bank, such as convenient locations, deposits by mail, convenient hours. ____

AUTO DEALERS

To sell specific merchandise. ____

New cars ____

Used cars ____

To sell automobile services. ____

Maintenance and repairs ____

To sell the broad image. ____

Reliability ____

Assortments ____

Good service ____

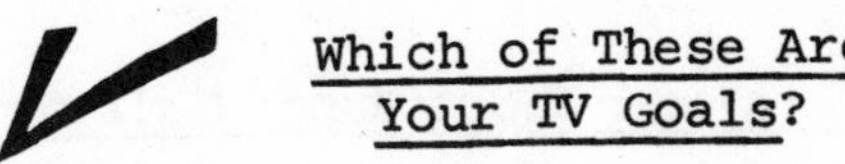

Typical Goals Which Stores Use on Television	Which of These Are Your TV Goals?
Promoting new departments or departments which have special potential. (These could vary from such areas as men's wear, women's fashions, hard and soft lines, special teen departments, college shops and so on.) List the ones that apply:	
______________________	______
______________________	______
______________________	______
Increasing brand name acceptance.	______
Increasing private brand acceptance.	______
Which of the store's private labels should be stressed?	
______________________	______
______________________	______
______________________	______
Increasing store traffic by item selling.	______
Using items which have proven successful in the store and building their success.	______
Using specially picked TV items.	______
Increasing store traffic by strengthening store events.	______
Store-wide events such as:	
______________________	______
______________________	______
______________________	______
Departmental events such as:	
______________________	______
______________________	______
______________________	______

Long and Short Range Television Goals

These are the store's goals translated into television goals. When completed, copies of these television goals should be in the hands of all executives concerned so there is a clear understanding as to what specific jobs television will do for the store.

LONG RANGE GOALS

1. Increasing acceptance of the store to give this image:

____

____

2. Increasing acceptance of these departments:

____

____

3. Increasing traffic through these kinds of promotions:

____

____

SHORT RANGE GOALS

1. ____

____

____

2. ____

____

____

3. ____

____

____

DECISION #3

Determine the Budget to Obtain These Goals

Advertising budgets vary from store to store. Of course, there are industry figures for different kinds of businesses that are issued by trade associations. These are norms or averages and represent large and small stores, profitable and unprofitable ones, all within the same sample. Samples include such retailers as department stores, furniture, drug, food, jewelry, and automotive dealers. Also, such local businesses as banks, insurance companies, funeral parlors, etc. These advertising statistics are splendid for comparing your own company with nation-wide averages, enabling you to determine why your company might be spending more or less than average.

Retailers' advertising budgets vary according to the characteristics of the retailers. For example:

- The well established store does not need as much advertising as the store starting out or trying to make a new impression on the community.
- Highly promotional stores with frequent sales need more advertising dollars than the "quality" type.
- Stores with prime locations do not need as much promotion as stores located in difficult places.
- High-ticket stores, such as furniture and jewelry, need more promotional expenditures. They do not attract as much traffic as a multi-departmental store. Their higher ticket sales enable them to have higher budget percentages.
- Local competition determines what a store may have to spend to "keep up with the Joneses."
- Stores with new branches or new departments require more advertising dollars.

Similarly, the store budgets spent on television vary from store to store. Some retailers spend little or nothing on television. Some spend close to 100% of their budget on television. In the past years, television budgets, as a percentage of total sales promotion budgets, have been climbing. This is true of the large chains that use the networks for national coverage of their stores as well as additional television in their local markets. This is true of the independent stores of all sizes.

In percentages, 20% of total advertising budget for television is a figure many stores use. Using this percentage, and a $10,000,000 retailer for example, the television budget might be organized this way:

Sales volume . . . $10,000,000 a year

The application of this principle can be applied to any store of any volume of sales.

Advertising budget . . . $350,000 a year

This is based on an industry figure of 3½% of store volume for this type of department store. This figure can be adjusted up or down depending on any store type. A store with a large amount of vendor co-op dollars might increase it substantially.

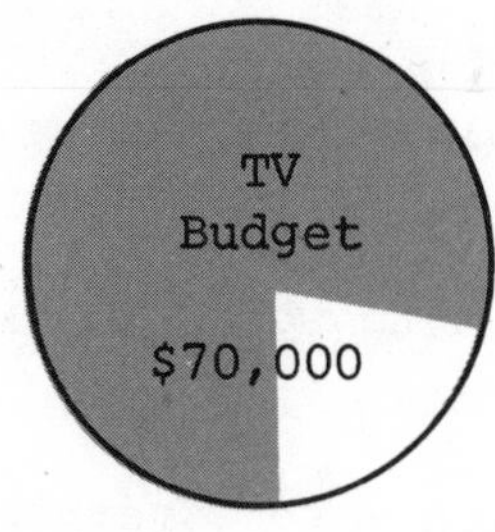

Television budget . . . $70,000 a year

This is based on 20% of the total advertising budget. The percentage might be higher or lower based on the goals of the store.

The next step in budgeting is to break the dollars down by month. In dividing the budget, the typical retailer distributes his dollars parallel to his sales volume or projected sales volume. Of course, in doing this, some dollars are "borrowed" from the peak months when traffic and sales are booming and added to the valley months where additional sales and traffic are badly needed.

In the budget chart, the monthly budgets are worked out for the $10,000,000 sample store. In the shaded areas, the spaces are left for you to work out your own budgets.

At the top, enter your own store volume, the total advertising budget for the year based on your own store's advertising percentage to sales. Next, indicate the television share of that budget based on your own determination.

In the example store, the monthly budgets are shown. The shaded columns are the spaces for you to work out your own budget.

Column A — The percentage of sales volume each month based on department store averages. If the store were not a department store, we would use its industry percentage.

Column AA — Adjust these percentages to your store's percentages per month, based on past records and projected goals.

Column B — These are the tentative television dollars for each month. Determined by multiplying annual TV budget of $70,000 by the percentage in Column A.

Column BB — These are your tentative television dollars for each month. Determined by multiplying your annual TV budget by the percentage in Column AA.

Column C — The monthly budgets are adjusted to add dollars to the low sales volume months, deducting these dollars from the high peak volume months. In peak sales volume months you do not need to spend the same advertising percentages because traffic flow is strong.

Column CC — Here you take your tentative monthly TV dollars in Column BB and round them out in the same way.

Column D — The minimum amount of television dollars each month is shown here. This becomes a steady base of dollars used each month.

Column DD — You adjust your base expenditure each month. By doing this, you create a basic 12-month base sufficient to keep your viewers aware of your store on a year-round basis. Also, it gives your television stations a knowledge of what you plan to do so they can anticipate your needs and work far ahead to obtain the schedules required.

Column E — This column indicates the additional dollars each month, except for the month(s) where only the base dollar amount is used. (These dollars are already accounted for in total annual TV budget.) Column E plus Column D equals Column C.

Column EE — In this last column you add your additional TV budgets per month except for the base month(s). Your Column EE should equal Column DD plus Column CC.

EXAMPLE:	YOUR STORE
Store annual volume.................$10,000,000 Total advertising budget at 3½%..... 350,000 Total TV budget, 20% of above....... 70,000	

MONTH	% VOLUME THIS MONTH	YOUR % VOLUME	TV BUDGET TENTATIVE	YOUR TV BUDGET TENTATIVE	TV BUDGET ROUNDED OUT	YOUR TV BUDGET ROUNDED OUT	BASIC MONTHLY TV	YOUR BASIC MONTHLY TV DOLLARS	ADDITIONAL TV OVER BASIC DOLLARS	YOUR ADDITIONAL TV DOLLARS OVER BASIC TV PER MONTH
	A	AA	B	BB	C	CC	D	DD	E	EE
JAN	6.5		4,550		5,000		5,000		--	
FEB	6.3		4,410		5,000		5,000		--	
MAR	7.2		5,040		5,000		5,000		--	
APR	7.1		4,970		5,000		5,000		--	
MAY	7.8		5,460		5,500		5,000		500	
JUNE	8.0		5,600		6,000		5,000		1,000	
JULY	7.4		5,180		5,500		5,000		500	
AUG	8.1		5,670		6,000		5,000		1,000	
SEPT	8.2		5,740		6,000		5,000		1,000	
OCT	8.5		5,950		6,000		5,000		1,000	
NOV	9.8		6,860		7,500		5,000		2,500	
DEC	15.1		10,570		7,500		5,000		2,500	
TOTAL	100	100	$70,000		$70,000		$60,000		$10,000	

This means the television budget over the 12-month period would look like this for the example store:

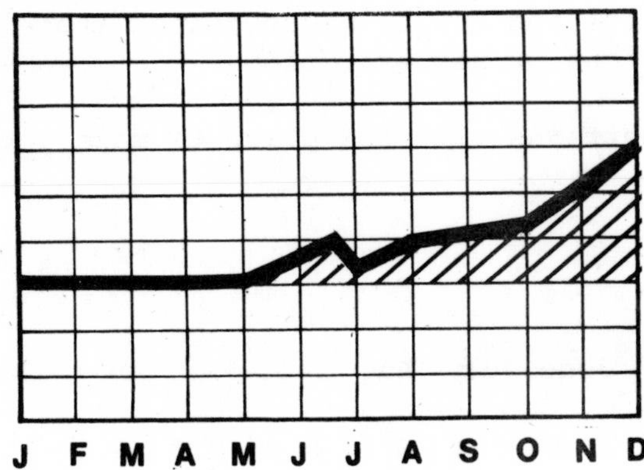

TOTAL TV BUDGET YEAR
. . . $70,000

ADDED BUDGET CURVE PER MONTH . . . on top of:

BASE BUDGET PER MONTH OF $5,000

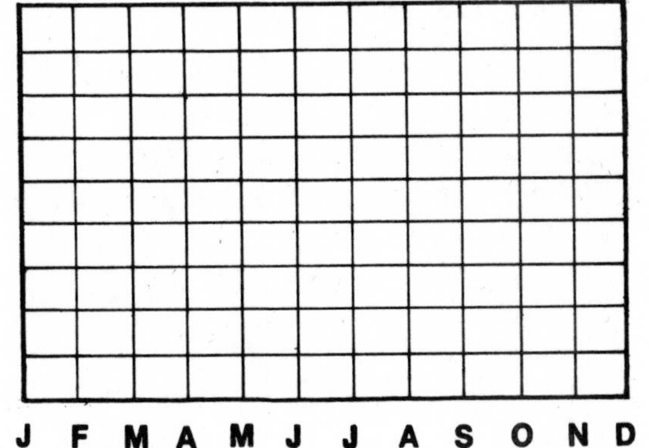

INDICATE THE BUDGET FOR YOUR STORE

TOTAL TV BUDGET YEAR
. . . $__________

ADDED BUDGET CURVE PER MONTH . . . on top of:

BASE BUDGET PER MONTH OF $__________

With this method of budgeting you are able to have:

- the steady or basic television advertising pressure each month to establish strong identity to your TV viewers.
- flexibility for your budget, applying added dollars in these months where you need it most.
- sufficient television dollars each month so your TV stations can adequately plan for your television campaign.

Percentage of Business Done Each Month Compared to the Total Annual Business

In the sample store used for the budgeting example, typical department store percentages of sales volume were shown. Industry averages exist for almost every industry. They can be obtained from the trade association of that industry. In examining these averages, however, a retailer adjusts his interpretation based on his own sales pattern local conditions and how he might change these sales patterns by new departments, new locations or new promotions.

Here are some typical industry sales patterns with the place to make your own adjustments.

	Jan	Feb	Mar	Apr	May	June	July	Aug	Sept	Oct	Nov	Dec	Total
Dept. Stores	6.5	6.3	7.2	7.1	7.8	8.0	7.4	8.1	8.2	8.5	9.8	15.1	100
Your Store													
Furniture	8.1	8.2	8.0	7.6	8.1	8.6	7.8	8.3	8.0	8.6	9.1	9.6	100
Your Store													
Jewelry	5.8	5.9	8.1	7.9	6.7	6.7	7.1	7.5	7.5	8.2	8.4	20.3	100
Your Store													
Men's Stores	8.0	6.3	6.9	7.4	7.9	8.1	7.0	7.3	7.6	8.4	9.5	15.6	100
Your Store													
Women's Wear	7.0	6.9	7.8	7.5	8.0	7.7	7.1	7.8	8.3	8.7	9.1	14.1	100
Your Store													
All Retail	7.9	7.6	7.9	7.7	8.3	8.5	8.1	8.3	8.2	8.5	8.7	10.3	100

March and April percentages change slightly from year to year depending on the date of Easter Sunday. An early Easter shifts some fashion sales from April to late March.

DECISION #4

Determine the People You Want to Reach

No store is in business to sell to everyone. The alert store decides what customers are the most logical for their particular type of store. Then it goes after these people.

When it comes to specific departments, especially for the items within the departments, the store knows, or should know, exactly what kind of people would be most interested in the departments and items. These people are the target audiences for the departments or specific items.

But for the store as a whole, you should determine what customers are the most logical and most important to the total store. The check list can help point the way.

✓ Profile of Our Best Customers

Sex

☐ Men ☐ Women ☐ If both, % to the total

Men_____% Women_____%

Age

Check the most important or number the sequence of importance. Sales Management "Survey of Buying Power" gives you the number of people in these age brackets in your market. Another source is your local Chamber of Commerce.

Age	Population In Our Store's Market
☐ 6 to 11	______________
☐ 18 plus	______________
☐ 18 to 34	______________
☐ 18 to 49	______________
☐ 25 to 49	______________
☐ 25 to 64	______________
☐ 50 plus	______________

Family Income

Check the most important or number the sequence in order of importance. Sales Management "Survey of Buying Power" gives you the number of families in each income bracket in your market. Another source is your local Chamber of Commerce.

Family Income	Number Of Families In Our Market
☐ Under $5,000	____________
☐ $5,000 to $9,999	____________
☐ $10,000 to $14,999	____________
☐ $15,000 to $25,999	____________
☐ $25,000 and over	____________

Where They Live

Check the most important or number the sequence in order of importance. Local sources give you the population of these segments in your market. They are also shown in Television Audience Survey books.

Location	Population
☐ City	________
☐ Metro or home county	________
☐ City & trading area	________
☐ Suburbs alone	________
☐ Other	________

Determine the Type of Customer for Each Department and Item

Before offering an item for sale, it is necessary to know who would buy it. This is obvious. However, many stores create advertising messages and insert them in print advertising or create commercials without asking questions about who would buy the merchandise:

- What sex? If both, in what percentage to each other?
- What age brackets?
- What income level?
- How can the store reach these specific kinds of people by media?

Some items appeal to a more general audience than others. However, most items appeal to a special audience. For the greatest economy in advertising expenditures, it is necessary to reach the target audience.

Appealing to Target Audiences With Direct Mail

Using direct mail, the store generally uses the customer charge list. These are the people who know the store. They demonstrated they like the store by shopping there and having charge accounts.

Many stores go another step and sub-divide their charge accounts:

- According to the departments in which various customers shop. These customers are reached with direct mail dealing with items in these departments or related departments.

- According to city and suburban customers. The customers are reached with items appropriate to their living habits and living needs.

- According to income levels. These customers are reached with items at appropriate prices.

Appealing to Target Audiences With Newspapers

Using newspapers, the store chooses the audience it wants:

- For adults generally -- the forward part of the paper.
- For young and middle aged men -- the sports pages.
- For women and men interested in finance -- the financial sections.
- For housewives and others -- the women's pages.

Separate sections have their appeal to different audiences. The alert store knows the audience it wants and studies the newspaper to decide where to place their advertising.

Appealing to Target Audiences With Radio

Using radio, the store has many choices. The program style of a radio station indicates the audience who listens to it and includes classical music, rock, middle-of-the-road music and all news and other variations. To reach the store's target audience with institutional messages or items, these are times of the day when various people listen most:

- A.M. drive time -- families at breakfast, men and women driving to work, teens.
- Housewife time -- late morning and afternoon reaching many women.
- P.M. drive time -- commuters on the way home, housewives at home, teens.
- Evening time -- family audiences in the earlier parts, adults only in latter part.

Appealing to Target Audiences With Television

Using television, the store has a medium that is researched in great depth. Television statistics tell you who is watching and when they are watching through the entire day. This information, available to every television advertiser, makes scientific selection of time slots possible.

First, of course, it is necessary to determine who the target audience is for each department and each item advertised.

TARGET AUDIENCE FOR TV PROMOTION 

People	Age	Where They Live
☐ Women	☐ 6 to 11	☐ City
☐ Men	☐ 18 +	☐ Suburbs
☐ Teens	☐ 18 to 34	☐ Both
___ Boys	☐ 18 to 49	**Income Level**
___ Girls	☐ 25 to 49	____________
___ Both	☐ 25 to 64	____________
	☐ 50 +	____________

Filling out this form is an essential step before:

- deciding when you can reach your target audience on television.
- creating your actual commercial appealing to your target audience.

DECISION #5

<u>Determine When You Can Reach These People</u>

Different people view television at different times. Actually, there is no mystery to the viewing habits of customers and potential customers. Once you know the living habits of people, you can determine their TV viewing patterns.

The viewing charts explain this. These charts are based on national averages. They are for Monday through Friday only. Not for weekends. Patterns can be different in your city if there are factories which operate at night, or if people go home for lunch.

1. Households Using Television

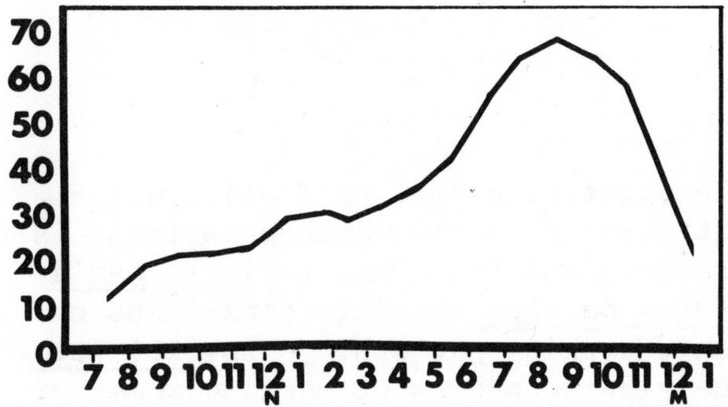

In this chart, the verticle line shows the percentage of people who view television. The horizontal line shows the broadcast day from sign-on to sign-off. The curve shows how customers watch TV when they wake up and throughout the day until they go to sleep again.

2. Television Costs Depend on TV "Circulation"

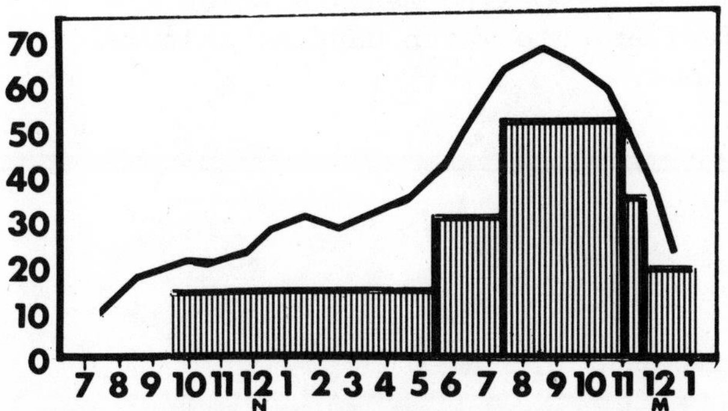

With two or more newspapers in a city, the one with the largest circulation costs more. After all, when you buy an advertising medium you are buying the attention of people. Television is priced the same way. The curve shows the same viewing curve as the first chart. The shaded areas show the relative differences in time changes. For example, prime time, attracting more viewers, is priced accordingly.

3. Women Viewers Compared To Total Household Viewing

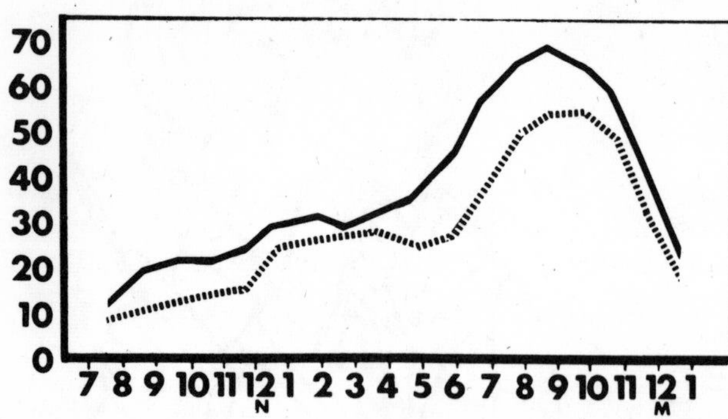

Women are the "purchasing agents for the family." Most stores view women as the primary target. The solid line shows the same viewing pattern for total households as the first chart. The broken line shows women alone. Housewife viewing is responsible for the large daytime TV viewing, especially early afternoon. For example, at 5:00 p.m. it is possible to reach almost 30% of the women in an average market on weekdays.

4. Television Costs For Women Viewing

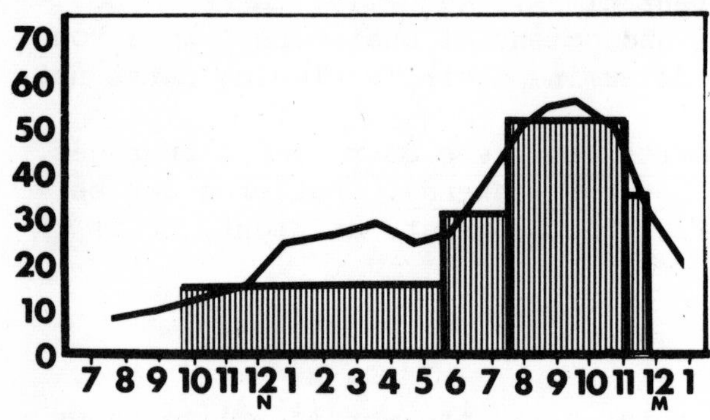

The viewing curve for women is far above the time cost bar during the afternoon hours. This makes daytime television an economical time to reach women.

5. Viewing Patterns For Men

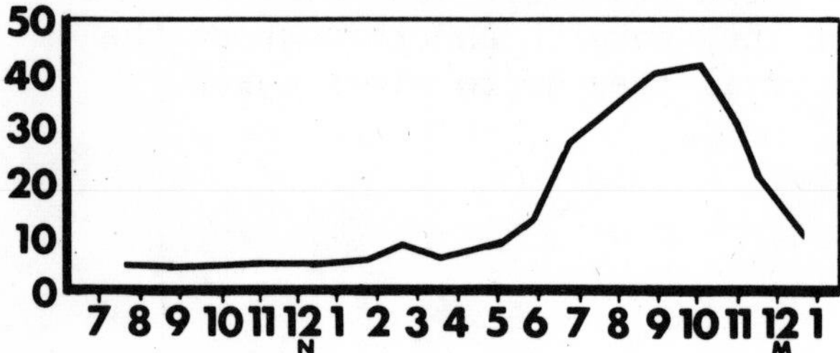

These patterns are, naturally, quite different from the women's patterns in previous charts. This reflects men's living pattern which is controlled by men's working pattern. A special note: the viewing habits of working women are almost the same as working men. To reach these workers, it is necessary to use evening time during work days. Of course, on the weekends, living patterns change. Working men and women are available to watch daytime television shows.

When You Can Reach Your Target Audience Based on Television's Basic Time Parts

As we have seen, television divides its day and charges for its advertising based on certain time parts. The time parts shown here are typical of the country. However, there are many locations where the hours indicated might be slightly different. It is necessary to check this with your own television stations.

TIME	EASTERN	CENTRAL	MOUNTAIN	PACIFIC		THE SHOWS	THE AUDIENCES
Day-time	Sign-on to 5:00 PM	Sign-on to 4:00 PM	Sign-on to 4:00 PM	Sign-on to 5:00 PM		Early News Children's Shows Panels and Games Local Shows	Adult men and women in early parts. Children. Housewives. Working adults on swing shifts.
Early Eve	5:00 PM to 7:30 PM	4:00 PM to 6:30 PM	4:00 PM to 5:30 PM	5:00 PM to 7:30 PM		Local News Movies Children's Shows Variety Shows Panels and Games	Housewives. Working adults in evening depending on local habits. Teens.
Prime Time	7:30 PM to 11:00 PM	6:30 PM to 10:00 PM	5:30 PM to 9:00 PM	7:30 PM to 11:00 PM		News Movies Various Entertainment "Specials" Documentaries Local Shows	All groups except young children in late hours. Teens.
Late Eve	11:00 PM to Sign-off	10:00 PM to Sign-off	9:00 PM to Sign-off	11:00 PM to Sign-off		News Variety Shows Talk Shows Movies	Adult men and women. Both working and non-working adults.

On weekends the audience for the time segments changes to match living habits. Teens are available throughout the day. Adults, especially men, will watch daytime sports shows.

6. Viewing Patterns For Teen Boys And Teen Girls

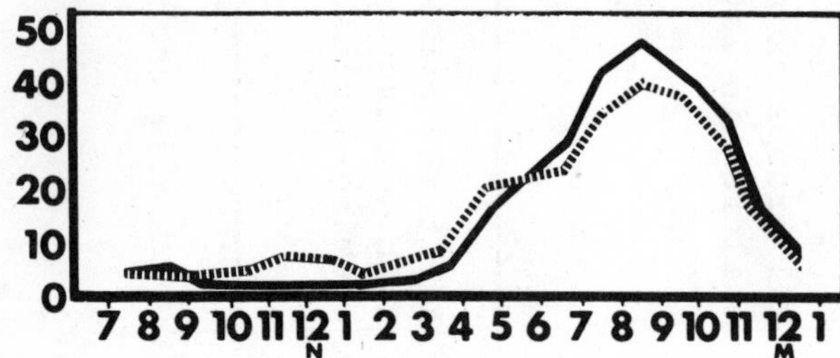

Once more we must think in terms of their week-day living patterns. During school hours they are not available to watch TV. But they are home in late afternoon and watch television. The solid line represents boys. The broken line is for girls who watch somewhat more than boys during the late afternoon then switch to slightly less after dinner time. With no school on weekends, the teen TV pattern changes. In summer months their daytime viewing also increases.

7. Weekly Viewing Compared To Daily Viewing

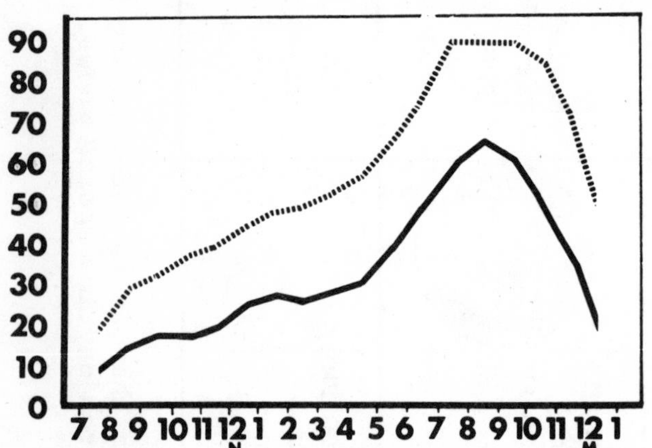

Obviously, not everybody watches television with same intensity every day. The broken line shows weekly viewing. It shows that within a given week practically everybody sees television. The broken line for weekly viewing almost parallels the daily viewing patterns...except for the higher percentages. The difference between daily viewing and total weekly viewing shows the need to run commercials throughout the week.

Remember -- All these charts refer to week-day viewing. On weekends, people spread spread their viewing through the day.

The time periods described are the basic periods. Often there are sub-[illegible]ions of these periods as any analysis of a station's rate card will show.

These periods attract different people in a way that different sections of newspaper reach different people. To reach men in a newspaper, the sports pages are useful. To reach men, and especially working men, television must think in terms of when they are available to watch TV. The charts show this.

DECISION #6

Determine What to Promote -- Events, Items, or Institutional Features

As indicated in the section on store goals, different stores take different routes on television. These routes, or goals, can shift from time to time. In fact, they can change overnight depending upon the circumstances indicated in the section on goals.

Here are television directions taken by some stores:

1. Build the store image

   This television approach is to sell the store on the image projected. Institutional advertising on television stands a better chance of being "read" than newspaper advertisements. Rich's in Atlanta is an example of a store which used television this way. Subsequently, it included the event and item route.

2. Build specific departments or areas of merchandise

   Some stores use TV to build a weak department. Better yet, some take a "hot" department and expand on its strengths. Alexander's in New York used television to expand acceptance for a shop appealing to a young audience. Hudson's in Detroit used television, tied in with all media, to build acceptance for different areas at a time. Men's Fashions. Women's Fashions. Home Lines. When the goals were achieved, television was used for other purposes.

3. Build a special "feeling" about the store

   A value store theme was used by Korvettes. They featured the star of "The Price Is Right" to promote the items. Garfinckel in Washington put over the feeling of a young-oriented, fashion-right store with the theme of "Your Garfinckel is Showing." They showed people in situations. No prices. At one time Neiman-Marcus in Dallas did a campaign to attract the younger, non-affluent customer.

4. Build traffic by selling events

   This is a favorite of many stores. Belk throughout the Southeast took its major events of the year. Television tied in by promoting these events and using items in the commercials.

5. Build sales of specific items

This is what most stores do in newspaper advertising. It is logical to do the same in television. Besides, by selling items of brand-name manufacturers, co-op advertising can be secured. While Sears and Penney, with stores from coast to coast, use network television as a "big umbrella," their individual stores generally promote items, usually with hard sell.

6. A combination of all the above

However, the typical store with a good television advertising budget will not go any one of the television routes. Instead, the store will use several approaches. Sometimes several at a time. Sometimes one and then shift to another one, when the goal of the first is accomplished.

SELLING ITEMS ON TELEVISION

In promoting items on television, the selling is best when it:

Demonstrates and shows items in use.
Has real people doing believable things.
Receives believable benefits.

Of course, a skillful TV writer can create TV advertising that hardly shows the merchandise itself. Instead, he projects the emotional benefits of owning or using the item or the service.

Items that do not sell without advertising will not sell with advertising. All the promotional skill available will not sell high-button shoes or starched collars or any out-of-fashion or unwanted goods. This is why stores feature best sellers, in-demand items and new items that they know will sell in substantial quantities and multiply store traffic.

Merchandise Timing Must Be Accurate on Television

Timing is of prime importance along with the item and its price. Timing, in television, is different than timing in print media. This is because of the nature of the medium itself.

People attend a Sunday football game and expect to read about it on the following day. They do not expect to find the news of the football game in magazines until a week after the game.

In television, this timing is unthinkable. The average viewer expects to see the game on television while it is happening. Seeing news at the same time it happens is taken for granted by viewers, especially during a football game or any news subject.

This feeling of immediacy has conditioned television viewers. In newspapers and direct mail, it is possible to offer items for a considerable period before actual need or use. In television, it is best to merchandise closer to the time the item will be sold. This follows for women's fashions, men's wear, garden tools and most everything.

As in storekeeping, there are many exceptions to this rule, such as unusual value offerings in fur coats ahead of the winter season, a price reason to buy slip covers ahead, a wide assortment reason to buy ahead for Mother's Day, June brides or graduates. However, the reason for buying ahead of the season must be explained to the television viewers as it would be explained in any other advertising medium.

Merchandise Selection for Television

Item selection for television should follow the same merchandising thinking used for promotion in any advertising medium:

- Items in demand, best sellers.
- Items that you know customers would want if they knew about them.
- Prices that are appropriate for the same audience. Avoid the top of the line and usually avoid the bottom price levels, too.
- Popular national brands or store private label brands that have local confidence.
- Promoted at the time of major acceptance. Exceptions -- as indicated before, off-season promotions are effective if a point can be made for the promotion. Example -- an air conditioner appliance in January because the store is holding a pre-inventory clearance.

Experience will show a store what items sell best for them. There are some stores which indicate that a carpet item or a mattress item is a sure hit. Some stores prefer to go the high-ticket route with applicances or furniture. Others believe in traffic items, such as hosiery, shirts, and slacks. There is enough evidence to show that all good promotable items sell on television.

TV Advertising Results

Advertising results vary from store to store. A store with a superlative reputation regularly gets greater sales results than a store in a less fortunate position.

One value-oriented store requires that each dollar spent in television produces $10 in plus-over normal sales of the advertised merchandise the next three days. Naturally, this applies only to in-demand sales items. For regularly priced goods, or areas of merchandise that do not have ready acceptance, the results would not be as impressive. In television, as in all advertising, item and price is not enough. An "institutional feel" about the store must be included in the message.

Scheduling TV Time for Items

Scheduling TV time for items depends on what the item is and what kind of results are wanted.

- for a sale or featured item where immediate sales are wanted, the schedule should be bought vertically. For example, if it is for Saturday selling, a heavy schedule should be run throughout Friday.
- for a brand name or regularly priced item where an impression of the item and its department is the goal, the TV schedule can be spread horizontally over several days.

SELLING EVENTS ON TELEVISION

Most stores promote several events each month. You might want to select your most important events each month and plan TV for them. They can be storewide, departmental, or a combination of related departments. Check last year's ads in newspapers and from the monthly sales promotion plans for this year. Prepare a monthly chart showing the date of the event and the general theme and specific items that were successful for the events.

In planning a major store event, heavy "saturation" can be used. One way is to use a large number of ten-second announcements to "sell the sale." With the repetition of these short announcements, it is possible to reach a huge audience of people. In print, you can kick off a sale with a full page, a double truck, or several pages. In TV, the repetition of short announcements gives the impact.

Depending on the length of your event, concentrate on scheduling the largest number of announcements one day before the event opens. Build up momentum with announcements and have merchandise ready when the sale begins. The 10-second commercials can serve as the "intro" for a full 30-second commercial. The added 20 seconds should focus on the items. Even if some viewers do not recall the specific merchandise, the repetition of the commercials sells the idea that there is much going on at the store and it would be a good idea to go down to the store to find out.

Typical Events for Television

As suggested before, step number one would be to make a 12-month list of regularly planned and used events, storewide, departmental and related departments. In the chart, the left column shows typical store events by month. The shaded areas show places where you can add your own events.

	Your Storewide Events	Your Departmental Events
January White Sales Storewide Clearances Fur Sales Furniture Sales		
February Furniture -- Home Furnishings Sale Valentine's Day Washington's Birthday Sale Housewares Events		
March Housewares Garden Supplies Spring Pre-Easter Promotion		
April Spring Cleaning Event Easter Fashions Garden Supplies Fur Storage		

May	Your Storewide Events	Your Departmental Events
Mother's Day Summer Sportswear Bridal Shop Outdoor Furniture and Garden Supplies		
June Graduation Gifts Father's Day Brides' Gifts Sportswear Vacation Needs		
July After July 4 Clearance Sporting Goods Furniture Sale (Last Week)		
August Furniture Sale Fur Sale Back-To-School Fall Fashions Fall Fabrics		

	Your Storewide Events	Your Departmental Events
<u>September</u> Back-To-School Men's and Boys' Wear Home Furnishings China and Glassware		
<u>September or October</u> Anniversary Sale		
<u>October</u> Women's Winter Fashions Men's and Boys' Wear Fashions, Accessories Columbus Day Sales Home Furnishings		
<u>November</u> Christmas Toy Promotion Pre-Christmas Values Thanksgiving Sales Home Furnishings		
<u>December</u> Christmas Campaign Gift Promotions After Christmas Clearance		

Scheduling Television Time for Events

Several points are to be considered in scheduling TV time for an event. (Length of the event and depth of merchandise are to be considered, too.)

If a real media mix is being used, you might "shoot the works" at the start of the event. Then let the other media carry the ball. Example: In an event that starts on a Thursday and goes all through the following week, "shooting the works" on TV could mean putting the entire television budget on Wednesday, Thursday and Friday.

If the TV budget is appropriate and the event is a week or ten days, you could open real strong, then coast with a minimum in between and come back with a strong impact for the last three days.

SELLING INSTITUTIONAL FEATURES AND SERVICES ON TELEVISION

Institutional advertisements in newspapers are often dreary affairs. As a personal medium, television can put over the emotion desired in a way print cannot.

Institutional television can range from a mood commercial showing fashion importance in a store to a person-to-person approach. A store head or a store buyer with on-camera ability can deliver a straight forward message with sincerity that is respected by the viewer.

Typical Institutional Themes and Services Promoted by Stores

Typical Institutional Themes	✓ Check the Ones Applicable for Your Store
Wide assortments	______
Liberal return policy	______
Famous brands	______
House brands	______
Competitive prices	______
Liberal charge features	______
Credit card	______
Top fashions for women	______
A "young" store	______
Teen shops	______
College shops	______
Leading men's fashions	______
Convenient store hours	______
Comparison shopping	______
Shop by phone or mail	______
Community services	______
Auditorium features	______
Charity causes	______
Conveniences such as restaurants, post office, baby sitters -- many others	______

Typical Service Themes	Check the Ones Applicable for Your Store
Reupholstering	________
Slip covers	________
Broadloom installation	________
Interior decorating service	________
Repair services such as watch, jewelry, appliances -- many others	________

Scheduling TV Time for Services and Institutional Themes

Most institutional themes and service-promotions are not geared to immediate response. Hence, vertical scheduling is not needed. Instead, it becomes best to spread the schedule over a period of days or weeks.

For the store that plans its promotional activity to include many sales, using institutional TV and service selling can very well be scheduled between the weeks of heavy sales promotion.

One highly promotional store ran an interesting experiment with reupholstering. In the first part of this experiment, they ran a half-page in the newspaper. It brought immediate and excellent response. It gave them more orders than they could handle quickly and properly. It jammed their workrooms and created a backlog of work.

In the second part of the experiment, they used their dollars on TV, but spaced their commercials over several days. Instead of a big rush of customers, the reaction was slow, but steady. Total response was as good or better than the newspaper approach. The advantage, however, was an even spacing of their workshop load so that customers did not have to wait for delivery.

DECISION #7

Securing Cooperation From Vendors

Manufacturers know the value of television. They put more of their advertising dollars into television than any other medium. It is no wonder that they offer co-op advertising funds for television.

Co-op dollars by manufacturers are a way of life. Manufacturers make them available to stores to promote their goods along with their regular advertising.

Manufacturers' co-op dollars are based on a variety of plans:

- A percentage of the store's own purchases. This percentage varies by manufacturer, from 2% to 5% of the orders placed by the store. Usually the store must match the manufacturer's money. The usual plan is 50%-50%. Some co-op plans are 75% manufacturer and 25% retailer.

- A percentage of the store's purchases on special lines of merchandise. Or, a manufacturer's plans for a seasonal period such as Christmas. Or, a plan based on specific kinds of merchandise the manufacturer wants to feature at a specific time. Or, the manufacturer might offer one kind of plan on a year-round basis and a different set of allowances for a seasonal or specific item. The matching of dollars, 50%-50%, might differ when the manufacturer has different plans.

- No percentage of co-op dollars based on store purchases at all. Although this is not frequent, there are cases where a manufacturer gives the store a free hand to use as much co-op not tied to purchases as the store wants. Or, a more likely arrangement where the manufacturer has one plan for his regular goods with the generous 100% allowance on some very special promotions.

- A fixed dollar allowance per item. Example: A $2 allowance on "X" brand mattress based on the number of mattresses purchased by the store from the manufacturer. In cases such as these, a manufacturer might have a year-round allowance plan for his regular line of merchandise and a different plan for his top-of-the-line brand. Or, as in other plans discussed, the different allowance might be for a seasonal promotion.

- A fixed dollar amount per case. This is a typical technique in the food and drug industry. Per case of goods bought from the manufacturer, the retailer gets "X" dollars allowance for advertising. The retailer must match these dollars, or not, based on the specific manufacturer's methods.

How the Co-op Dollars Are Rounded Up

The individual store buyer is the key to obtaining co-op funds from vendors. Many stores consider this job only secondary to the actual buying of goods itself. Many stores insist that the buyer ask for co-op funds, or an update on the manufacturer's co-op plan, every time an order is placed.

How Stores Use Their Co-op Funds

Different stores use co-op funds in different ways. Or, they may use a combination of the plans.

Ways Stores Use Co-op Dollars As Part of Their Total Promotion	✓ Check Performance for Your Own Store
A year-round plan to sell major brand names and establish the store as the place to go for the top brands.	______
Items promotion within the regular schedule of the store's advertising. In these cases, co-op is used throughout the year.	______
Event promotion or "shooting the works" during key promotions. Example: Saving co-op funds for sheets, cases, etc., to use during a January White Sale and other white sales such as in May. Example: Saving co-op funds for furniture and bedding and using these dollars to "back up" the February and August Furniture sale. Example: Bulking co-op money for Christmas, or Father's Day, or Mother's Day, or Anniversary Sale.	______
Themed promotions by which a store promotes:	
• a basic style feature throughout many departments.	______
• a new fiber or design.	______
• a whole area of the store, such as the fashion departments.	______
• the men's clothing, or the children's departments, especially back-to-school periods.	______

The latter technique is used by stores as all-out promotions. They usually combine all media available, with TV playing an important part.

Co-op Items Tied Into a Theme

The theme of the promotion is used in newspaper captions and newspaper institutional advertising. With television, a combination of television commercials can be used with:

- A commercial of 10 or 30 seconds which just sells the theme.
- A commercial of 30 seconds with the opening 7 to 10 seconds selling the theme, and the balance selling a specific item. The item is one which carries co-op dollars.

------------------------30 seconds------------------------

THEME OF PROMOTION	Manufacturer's item integrated into commercial. Store "sig" at end.

Repetition of the theme from commercial to commercial or to many different manufacturers and products sells the theme quickly and solidly on TV.

Selling the Store's Co-op Plan to Vendors

As indicated before, buyers have the job of getting co-op funds. To bolster buyers, however, many stores create various TV "packages." These packages show a manufacturer what he can get for his co-op dollars.

The package plans are presented in many ways. They can be as simple as a mimeoed sheet giving details on costs. They can be elaborate brochures. The brochures would tell the story of:

- The store's market and why the market is important to the manufacturer.
- The store's goals in selling this market.
- The various media the store uses. Newspapers. Radio. Direct Mail. Examples are shown.
- The television plans of the store for year-round, or a special promotion.
- The "package plans" would show various combinations of what you can get for $500, $1,000, $2,000 and so on. The buyer would usually show the manufacturer the plan that would be most appropriate.

A sample page of package plans is shown here.

Production Costs and Vendor Participation in Them

Based on what TV commercials cost a store to produce, many retailers establish a rate for preparing their co-op commercials. They ask the vendors to share in the costs of the TV commercial on a 50%-50% basis in the same way as many stores ask production costs for co-op newspaper ads.

Where a manufacturer has an acceptable film and the film can be used with some changes, the store's production costs are lower and the store adjusts its price.

There are cases where the store offers to become the production facility for the manufacturer of certain co-op commercials. In these cases, stores feel that they are entitled to the full cost of the TV commercial and the manufacturer is able to use it in any other market.

Whatever TV production method is used by the store, a team is needed to put together the plans for the manufacturer, for the package plans, for the production costs, and for the manufacturer/store contract. This form can act as a guide to the store team:

Example of a "package plan": The local advertiser might have several plans at different dollar amounts starting as low as the store felt needed and going to higher cost packages based on the different vendors.

(STORE NAME)

TYPICAL WEEKLY TV SCHEDULE -- 40 SPOTS

30 SECONDS -- MONDAY THRU FRIDAY

Time	Spots Per Station WAAA	Spots Per Station WBBB	Total Time Charges	Combined Rating Data and Audience: Gross Rating Points	TV Homes (000)	Total Women (000)	Total Men (000)
Day (9:00 AM - 5:00 PM)	6	5	$ 600	110	480	387	133
Early Eve (5:00 - 6:30 PM)	5	5	1,000	140	683	504	443
Prime (7:00 - 10:00 PM)	5	5	2,500	220	1,010	844	604
Late (10:00 - 10:30 PM)	4	5	1,000	125	469	358	315
Weekly Totals	20	20	$5,100	595	2,642	2,093	1,495

Ratings and audience based on most recent rating book. Details available from (Store Name).

Total Time Charges	100%	$5,100
Vendor Time Charge Share at	50%	$2,550

Example of an agreement between the manufacturer and the store:

____

(Store Name)

Date____________________

This is to indicate our plans to participate in a television promotion with you.

The merchandise to be promoted will be:________________________________

____

____

and all details of the merchandise, prices, etc., will be furnished by us to you.

The TV package plan to be used is your plan____________________________

____

for $____________ based on our co-op plan of__________________________

____

Upon completion of the campaign, you will furnish us with a bill plus proof of performance and copies of the script used.

____
Manufacturer's Name

____
And Representative

____
Store Representative's Name

Co-Op Planning Procedure Check List

Steps Needed	Who Will Do It?
1. Different Package Plans based on different total prices.	______
2. Production Costs for commercials. (Estimates to be secured from stations or production houses.)	______
3. Writing agreement between store and manufacturers.	______
Store legal department clearance.	______

In the actual production of the commercials, there are usually two possibilities:

- Each commercial produced as a separate unit with the features brought out accordingly. However, a store "family relationship" in appearance is used from commercial to commercial.
- The commercials become part of a "sandwich" or "doughnut" campaign with considerable similarity from commercial to commercial.

-------------------------30 Seconds---------------------

Intro selling the store, a feature of the store, a store event.	Specific manufacturer's segment of the commercial and a store "sig."
---7 to 10 Seconds--	----------20 to 23 Seconds----------

Presenting the Total Plan to Store Buyers

After the specific store-wide vendor plan is created, it is generally presented to the store buyers in a regular meeting. Store head, merchandise manager, and sales promotion manager kick off the meeting and:

- Explain co-op policies of the store.
- Explain co-op relationship in TV.
- Explain the store's coming campaign in which TV becomes an important part.
- Hand out copies of the co-op plan and "packages." Explanations on how to "sell" them.
- Hand out the agreements to be signed by the vendors.

Expanding the Co-op Plan With Vendor Meetings

Many stores have "dressed up" their proposals to vendors with considerable television dramatics. Often the store will enlist the television station to help them produce a co-op presentation on video tape.

The presentation on tape or film usually follows this sequence:

- Store head addresses himself to the vendors. Tells about the store, background, and how to cooperate with its resources to sell more of their merchandise.
- Sales promotion executive picks up and tells how TV is an important part of their media mix.
- Station announcer, because he is at home in front of the camera, discusses the store's market and the need to reach the entire market. He shows what the store is doing in print, on radio, on TV, and how it plans to do more on TV. He rolls "spec" commercials the store will use of typical manufacturer's items.
- Co-op plan is discussed by a store person, president, sales promotion or merchandise manager.

The tape or film presentation would be shown at a buyers "kick off" meeting.

The major use, however, would be in meetings on key markets where vendors are invited to hear and see the store's marketing plans. Naturally, follow-up sessions are used by the buyers who call on these vendors. They take their information on the store's plans in addition to the contract agreements.

DECISION #8

Determining Who Does the Television Advertising

If the television advertising job is everybody's job, nothing really gets accomplished. Depending upon the size of the store and the size of the television budget, a store's television advertising staff might be a staff of many people, or it might be the part-time activity of just one person.

Regardless of the size of the television advertising organization, the duties are the same. They include management, budgeting, graphics, writing, display props; production, media buying, and evaluation.

The Role of an Advertising Agency

Some stores rely on an advertising agency to give them complete professionalism. In this case, the advertising agency becomes an extension of the store's own advertising department. It is, or should be, involved right from the start in store sales promotional planning. In this way it can fully interpret the store goals for the proper use of television needed to achieve those goals. They can blend the store's television with the entire media mix.

Some stores limit the use of their advertising agency to production and time-buying. In these cases, the store can easily short-change itself out of the multi-talents that an agency can give them. Where the store limits the agency activity, the store gives the agency the events and items it plans to promote and tells the agency to produce the commercials and buy the TV schedules for them.

The Stores That Do the TV Job "In House"

Other stores tend to do the entire job themselves. These stores feel that since they create their own newspaper advertising, they should be able to create their own television advertising. Most stores in this category use the studio facilities of a television production house or their own television station. In the case of the production house, there are varieties of reliance. Some stores will turn over a script and a storyboard and say, "This is the way I want it. Go make me a commercial." Other stores will give the production house the rough idea and ask that they improve on it and do the production with or without store supervision.

Some stores actually do the entire job themselves:

- the budgeting and scheduling.
- script writing, by the advertising department copywriters.
- storyboards, all graphics and supers by their artists.
- props by their display department.
- advertising department photography studio becomes the television commercial studio.
- camera people are on staff or free lance.
- models are on staff, possibly those used for fashion shows, or from model agencies.
- audio voice by their own people or free lance talent, possibly from a radio station.
- finishing, both developing negatives and final positive prints by commercial houses.

There are all kinds of variations of the store-controlled television creation. If a store is big enough and the television activity is large enough, the store might consider doing most of the above functions, but leasing the facilities of a good local production studio, which could be a TV station, is better than attempting to create and use ineffective in-store studio facilities. In leasing a studio, you usually obtain the full crew as part of the facilities. This may include a capable director as well. Fortunately, many stores take this latter step of leasing a top-notch studio and its professional crew. In the long run it may be cheaper.

Regardless of the store size and television budget size, the store executive who heads the television activity should be a promotion-minded merchandiser or a merchandise-minded promoter. He need not be a technical person in the same way as the retail newspaper advertising executive.

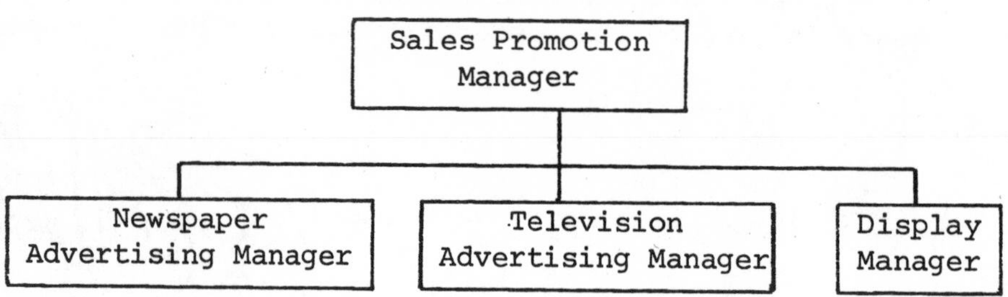

In establishing the television-advertising department, whether it is one person or many, these are the duties which it should exercise:

Major Duties of the Television Advertising Department or Unit	✓	Who Will Do This in Our Store?
Interpret the store goals and its advertising goals into television goals.		________
Establish the television budget in cooperation with the sales promotion head of the store and the controller.		________
Coordinate the store's promotional themes, storewide events and major departmental events into the television campaign.		________
Work with buyers, merchandise managers, and the advertising departments to decide which items to promote.		________
Write, or direct the writing of, scripts on the items for television. The scripts must reflect the store's character.		________
Work with the store's advertising artists for needed art work, including storyboards and supers.		________
Work with the store's display department on props for the commercials.		________
Deliver the merchandise and props to the studio according to schedule. Check the items upon return to see that they are returned to stock.		________
Follow through with store promotion department for:		
(a) Proper "signing" of items advertised.		________
(b) Good displays in the department.		________
(c) Displays of related or higher quality items.		________
(d) Windows and other tie-ins.		________
(e) Drop-in advertisements and separate advertisements to promote the television.		________
(f) Direct mail enclosures, handbills, stuffers, and charge statement enclosures.		________

Who Will Do This in Our Store?

Buy the television time using professional aids, such as the rating services. ________

Keep the store's personnel informed:

(a) Mimeographed bulletins to sales people telling them of television-advertised items. ________

(b) Bulletins to floor managers, mail and telephone order departments, and store executives about the television-advertised items. ________

Checking all television bills, including time charges from stations to see if all commercials ran as scheduled or proper make-goods were used. ________

Keep records of television results to enable repeats of best selling items. Report this information to proper store executives. ________

Conduct store and department studies to decide suitable departments and items to promote. ________

Study television methods of stores and other advertisers to learn improved techniques for store selling. ________

Maintain continuous contact with the television stations and the store's advertising agency, if one is used. ________

These television duties show that the store television head is more than a sales promotion or a merchandising executive. He is a combination of both. It is best for him to be located in the store's advertising department where he reports to its head. In large stores he may have several assistants, and be on a par with the advertising manager who works on printed media. In smaller stores, he would do other activities besides television. Regardless of how store television is handled, the complete responsibilities and duties for television should be delegated to one person, if maximum results are to be obtained.

DECISION #9

Writing the Commercial

Print advertising, as indicated before, has been defined as "salesmanship in print." Television has been defined as "personal selling on a mass basis."

Today, almost every type of store uses self-service or at least self-selection. This is true, not just for supermarkets, most variety chains, discounters, drug stores and automotive supplies, but also for department and specialty stores.

This change in store selling was made possible by advertising which pre-sells goods not only by the paid media advertising of newspapers, radio and television, but also by windows, interior and point-of-sale displays. In other words, advertising, to a large extent, has replaced personal selling and this will continue to be the trend.

Writing advertising, whether it is for television, radio or newspapers, is unique and the finished product is not meant to be a literary masterpiece. Instead, the writer takes the role of the salesman and motivates the customer to the cash register. Direct sales returns are the usual test of whether a piece of copy is effective or ineffective.

Copywriting begins with gathering the facts about what is to be sold, whether it is an item, an event, a department or a whole store or specific services within the store. There are several places to learn about the merchandise:

Accumulating the Facts About the Merchandise

Step #1 -- The Buyer. The buyer who originally bought the merchandise should be able to explain why she bought it in the first place. The buyer can give such details as:

- The merchandise's main asset from the customer's point of view.
- Secondary benefits from the customer's point of view.
- Fashion features.
- Colors and sizes
- Performance, washability and construction.
- Economy and quality details.

Step #2 -- The Salesperson. This is the person who sells the goods. The salesperson, if she knows her customers, should be able to give many ideas:

- Why she thinks the item is good from the customer's standpoint.
- The one big feature which she would focus on in her first contact with the customer.
- What kinds of customers would buy the item -- sex, age, income level of customers.

Step #3 -- The Customers. Nothing is as advantageous as talking to prospective customers to learn about the merchandise itself. Find out why she would buy it:

- Is it for the fashion features?
- Is it for the overall quality?
- Is it for durability, washability, or something about the construction?
- What is the most useful aspect of the item?
- What kind of people would but it? -- sex, age, income level.

Step #4 -- Where the Customers Live. Living conditions in the city or suburbs or both can determine the style of the writing as well as which features to stress.

Step #5 -- When It Is to Be Sold. The time of the year will determine specific features for emphasis. This would be true of clothing, winter versus summer weights of materials, various appliances, furniture for outdoor versus indoor types.

Step #6 -- Make an Inventory. Write down all the facts. Try to think in terms of the main benefits that the customer wants in the goods. Think in terms of what person is the most logical customer for it.

Example: If the item is a man's topcoat, the inventory would start:

TOPCOAT	CUSTOMER
Selling Price: $62.50.	Salary averages $150 a week.
Regular Price: $95.00.	Cannot afford more than $95 for a coat.
Hard finished woolens.	Wears coat every day -- business, sports, dress.
Hand tailored details.	Must stand up for four years.
Long-wearing.	Sizes to appeal to every man.
Used in a short season.	Styles to appeal to every man.
Does not wrinkle easily.	Lives in immediate area and suburbs.
Rainproof.	Able to wear it with sports or business suits.
Choice of colors -- list them.	
Choice of styles -- list them.	
Choice of sizes -- list them.	

Step #7 -- Select the Most Important Benefit. The customer is only interested in the item you have to sell from her standpoint, not the store's. Knowing the merchandise, the next step is to think in terms of the greatest appeal the item has for the customer. Customer benefits can include:

- MONEY -- You can tell and show how they can save money.

- POPULARITY -- You can tell and show how the item will help achieve popularity. Show this benefit in items such as sports equipment, bars, games.

- FASHION LEADER -- You can tell and show how the item sets the customer apart as a fashion leader. Show this benefit in new styles of clothing.

- BEAUTY -- You can tell and show how the item makes the prospect more attractive. Items such as beauty preparations and clothes benefit from this approach.

- COMFORT -- You can tell and show how the item puts the customer at ease and makes him comfortable. Furniture, shoes, sleeping equipment are examples.

- FUN -- You can tell and show how the item adds to enjoyment. Use this benefit for games, sports clothes, and other items.

- LEISURE -- You can tell and show how you achieve the benefit of more time in washing machines, automatic toasters, and other appliances.

- LOVE OR PRAISE -- You can tell and show how the item will increase attention to the customer and help him receive praise. Certain clothes, cosmetics, anti-perspirant preparations and, often, books find this selling benefit important.

- HEALTH -- You can tell and show how the benefit adds to health. Drugs, vitamins, sunray lamps, swim suits, and some clothes can be sold this way.

- LEADERSHIP -- People want to be leaders. Tell and show how the item creates the leadership benefit. Items such as television, new fashions and new gadgets of every kind.

- ADVANCEMENT -- You can tell and show how the item will help the customer get ahead. This benefit will sell educational books, some clothes, fine furniture.

Appeal To the Five Basic Senses

In examining possible benefits to feature, it is important to appeal to the five basic senses. In print advertising, the only senses reached automatically is the one of sight. The others must be reached in the copy. In television, you appeal to both sight and sound. The other senses must be appealed to in different ways:

Touching or Feeling	Stroke a cashmere sweater, a soft towel or blanket while you are describing it to portray the feel.
Smelling	Have the model smell the frangrance of perfume or the aroma of good food while you are describing it.
Tasting	Have the performer eat the delicious candy or cakes or other food products while you are describing it.

The Four Basic Parts in Print Advertising

Print advertising primarily divides into four components:

- The headline.
- The picture.
- The details.
- Where to buy it.

The Headline. This is the most important benefit of the merchandise from the standpoint of the customer. Often a subhead is used. In this case, the subhead could amplify the major benefit or give additional uses.

The Picture. In regular merchandise advertising, the picture conveys how the merchandise would be used. Of course, in some forms of advertising, the picture is used simply as a "stopper" to attract the reader to the page.

The Details. This is always an expansion of the main benefit of the merchandise. Other sub-benefits can then be featured along with details of sizes, colors, styles. Price becomes an important part along with comparative prices if the item is on sale. A fresh, news approach without tricky words is best. Short. Concise.

Where To Buy It. This is the store signature and the addresses. Store hours might be important as well as other features such as parking. Credit terms might be added if the item is large ticket.

The Three Basic Parts in Television Advertising

In print advertising today, the bookish copywriter has gone out of style. Instead, she practically visualizes the advertisement in print. If the copywriter is not a good visualizer, she will meet with the graphics person to explain her key idea for the advertisement and get a visual play-back from the art person. A merger of thoughts from the two finally evolve into the finished advertisement.

In television, it is essential to think of audio and video at the same time. In the typical commercial, this is developed into three parts:

The Opening	The Middle or "Sell"	The Close

The Opening or "Intro"

There are two ways to open a retail commercial:

(1) Tell something about the store -- what it stands for, or the special sale that is being held.

(2) Open with a surprise idea aimed to hold attention.

The opening technique depends on the situation and the store. For example, a store known for big values might open with its name because it is identified with value. A high fashion store might do the same. The name is really saying, "Look at me...I am the most important fashion store in this city...so pay attention if you want to know what is good for you." On the other hand, a store with little fashion acceptance would turn people off if it blasted its name at the start of the commercial of fine fashions. In this case, the store would do best by coming on with the wonderful fashions, selling them on the air and then telling where to purchase them.

The Middle or Sell

This is where the merchandise is shown, demonstrated and sold. The one big benefit is shown and explained. If there is time, other benefits are explained. For example, if the item is a man's suit which can be worn and packed with a minimum of wrinkles and if this is the main benefit, it becomes the key to the commercial. Show the suit in action; bending, twisting, working, or walking or playing tennis. If there is time, then go on to explain the fine tailoring, the many fabrics and other great features. This is the time to show the price. Show it with supers to have the price register with the eye and ear.

The Closing

This is where you ask for the order. A good idea is to repeat the big feature, both in the video previously used and in the audio. Then indicate where the merchandise can be bought.

Print and Television Writing Comparisons

Time is the big factor in television. In print, regardless of the size of the space, it is possible to use smaller type to squeeze in a message. With television there are only so many seconds. There is no such think as rubber type:

- 30 second commercials may only take 60-65 words.

- 10 seconds may only take 20-25 words and even less if it is an ID (Station Identification) which means a few seconds must be given up to show station call letters.

Things to Watch for in Writing Television Advertising

It is good to remember the country parson who told his formula for a Sunday sermon. First, I am going to tell you what I am going to talk about. Second, I'll tell you. Third, I'll tell you what I talked about. In television writing this would be: (1) Tell. (2) Re-tell and show. (3) Re-tell and show. (4) Tell where to buy it.

In other words, with time so limited, it is good to zero in on certain essentials and not confuse the viewer with unnecessary details.

1. Find and talk about the one big benefit. Do not give a half-dozen reasons why the viewer should buy the item. Think about stopping at the sales counter where the merchandise is located. What would the expert salesperson say to you if he had split-seconds to stop you and interest you in the item? That is the big benefit to use on television.

2. Show the dominant graphic idea. It should parallel the dominant benefit (and repeat, if possible, the same audio and video benefit later in the commercial, even if abbreviated).

3. Show what you are talking about. Coordinate. Never talk about one thing and show another.

4. Add impact to the benefit. Show the key words in a super while you talk about them; let the eye and ear work together on the main point.

5. Get in tight, tight, tight. This is no place to kid around. If a benefit can be shown, remember the TV screen is small. Show the collar line, the fabric detail, the automatic switch and show the close-up detail.

6. K.I.S.S. That magic word which means "Keep It Simple, Stupid." Eliminate all unnecessary words. Eliminate all unnecessary details in the graphics. Think of the display people in the store and how they keep window displays down to bare essentials to get over the major points.

Print Copywriting Versus Television Script Writing

For television, copy is written as script. This is similar to the writing formats for plays. The script shows two kinds of action. They are side by side:

One side is the video column. It tells what the viewer will see. It includes instructions for the camera and the lighting. It should be detailed enough to give understandable directions.

One side is the audio column. It tells what the viewer will hear and includes instructions for the music. Inflection and pace of the voice is explained.

Both together, side-by-side, are designated as a frame. If subsequently translated into a storyboard, the boards will show the separate frames.

The instructions in the script are given in technical abbreviations as the script samples show. For example:

LS -- indicates a long shot.
MS -- a medium or bust shot.
CU -- close up.
CUTS -- indicate a quick transition from one scene to another.
DISSOLVE -- indicates the slow transition with one scene fading out while the other appears.
MU -- means music under when the voice comes out stronger. Sometimes referred to as down under.
VU -- can mean that the music is heard above the voice.
VOICE ON -- the performers do the talking.
VOICE OVER -- the performers do not talk, but the voice is heard.

Television's Quick Messages Require the "5-S" Approach

The key to successful TV commercials is to make them easy to understand, remembered, and acted upon. The "5-S" approach should be used in both video and audio.

	VIDEO	AUDIO
SIMPLICITY	Easy for the eye to see.	Easy for the ear to hear.
	Simple sets, uncluttered and to the point.	Uncomplicated and understandable words.

These are achieved by using one basic benefit and featuring it in video and audio. Together, the effect should remain in the viewer's mind. Avoid fast movements or frequent cuts or many dissolves which bother the viewer.

	VIDEO	AUDIO
SINCERITY	Pictures that are in every day reality.	Words that are in every day conversation.
	People who look real for the part they are playing.	People sounding like the people they are playing.

These are achieved by keeping to things that are believable. You cannot use a Hollywood star type in a supermarket set or a truck driver type in an expensive men's suit ad. The typical way an item is used or worn by the usual kind of people who use it is best. The usual way an item performs is the best way to portray it. If an item performs in more than one way, such as a sofa-bed, show it.

	VIDEO	AUDIO
SELL-ABILITY	Prove the fashion, the value or the price with the picture that best portrays this.	Prove the fashion, the value or the price with words that give the benefit reasons.

And make the sell-ability do double duty by using supers on the key points.

	VIDEO	AUDIO
SHOCK	Picture something unusual about the item.	Say something unusual about the item, if possible.

A sofa which opens to a bed. A suit where the jacket can be work with odd pants. A watch that looks like a pretty bracelet. A night gown that is pretty enough to be an evening dress.

	VIDEO	AUDIO
SOLD WHERE	Show the store signature in full screen. If the "sig" is a fussy one, simplify it slightly.	Say the store name in the commercial several times and especially when it is shown on screen.

Using Print Knowledge for Television

The person already in advertising, who has prepared newspaper advertising, has a head start in creating television commercials. This is especially true for television meant to sell items or events or a combination of the two.

Using print advertising knowledge and making the translation to television is effective for simple, inexpensive commercials. Many are done in this way. There are big differences in newspaper and TV copy, but the knowledge of print can place the copywriter into the "advanced class" of TV almost at once. Subsequently, with practical, in-studio background, it is possible to graduate from the high school class of television to post graduate work and be able to create original, exciting commercials. Here are the steps in print to TV translations:

Writing the Print Advertisement

1. Write the copy for an item as if it were going to be a newspaper ad. Write a selling headline promoting the major benefit of the item or service.

2. Write a sub-head to further develop the headline. Then support the value with price or comparisons.

3. Write a copy-block spelling out the customer benefits of the headline plus additional facts.

4. Finally, the request for action. Where? When?

5. Show the store signature.

6. Make a layout of the advertisement. Use your knowledge of graphics.

Start to Translate the Print Advertisement -- Radio

7. Talk about the merchandise. Pretend you are on the telephone to convince someone about the merits of the item. Take these steps:

   A. Speak in the first person, conversational style. Use spoken words rather than written words. Picturesque slang. If the item is one which appeals to the younger generation, use their language.

   B. When talking about the merchandise, you would actually be doing radio copy. In an imaginary telephone conversation, think about what the other person might be saying to you about the item or questions being asked. In this way, you can project the copy idea to a radio (and later a television) commercial where the two performers are doing the commercial. Examples: Housewife and friend. Salesman and customer. Two men friends. Two children.

8. Speak face to face to a customer. You are the salesman. Your style of talking changes. You find yourself showing the merchandise, and demonstrating the merchandise.

Continue to Translate the Print Ad -- Television

9. Think in terms of the merchandise displays in the store. How would your display department show the merchandise in the store? What accessories? Suppose you were showing a raincoat. Would he have rain dripping on the coat? Would his model wear rubbers? A rain hat?

10. Re-cap! You wrote a print advertisement. You used this knowledge to turn it into a radio advertisement. You added the demonstration qualities of the salesman and the display man. You are on the way to TV copy.

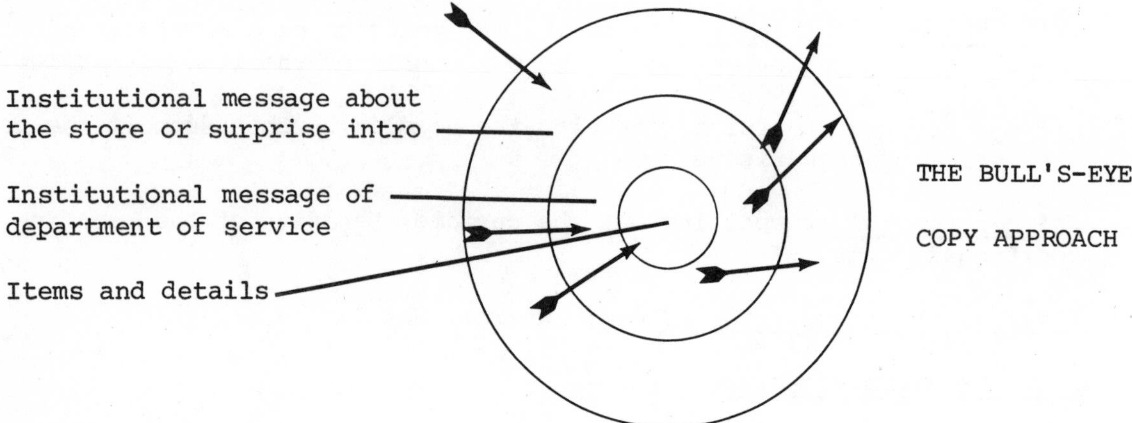

Think of the bull's eye shown. In the bull's eye technique, you start with the outer area with copy that does an institutional sell of the store. If your customer is not interested in the item, at least you have made an impression about the store. (Of course, there are times when you start with a surprise intro.)

11. Move closer to the item and institutionalize the department or the general classification of the merchandise.

12. Hit the bull's eye. This is where you hit the item. After you discuss the item and its price, you go right back to the middle circle, institutionalize the item and the department once again, or possibly discuss a credit feature, a service feature: credit, easy parking, or some other service. Finally, you move to the outer rim where you conclude by institutionalizing the store.

A Finished Commercial -- But Only a Start

Easy? Simple? For "meat and potatoes" television selling, this approach would do the trick. Many commercials are created this way.

But remember, these are only starting points and rules to get you going. When you are deeply involved in television, you can break the rules completely to accomplish many things.

In this "translation" of print to TV, no reference was made to music. Music can be a background to tie all of your commercials together. It may be a jingle which is used over and over again. In radio. In store public address system. In fashion shows. Everywhere, so it becomes your own theme song.

More on Translating a Print Ad Into a Television Commercial

Television Bureau of Advertising (TvB) is the industry organization which promotes the use of television as an advertising medium. It is supported by stations, networks and station representatives. TvB involves itself in wide areas of research. It reports on TV case histories in different areas of business. These assist advertisers and show them what others have done successfully. A large part of their work is educational. This includes assisting advertisers on how to use the medium.

TvB invented the "print converter." By the use of transparencies that have cut-outs resembling a television screen, TvB suggests how it is possible to take a newspaper advertisement, select its basic components, and turn them into a workable television commercial.

Their example shown here is of a mythical store called Noble's and their newspaper layout on a sale of television sets.

Using the print-to-television discussion in the previous section, think in terms of the bull's eye approach. See how the most important elements of the advertisement have been indicated with television screens and how they are converted into a television advertisement.

To produce the Noble advertisement as a television commercial:

- Original artwork and glossies of the type could be assembled.
- Stats or photos of all artwork shot to be the same relative sizes. Preferably to fit a 14 inch by 11 inch art card.
- Each frame would be mounted separately on gray stock.
- Conflicting type and other elements painted out and eliminated.
- Color added in the backgrounds and on the sets themselves.
- Store signatures retouched to give color.
- The sequence of the art determined.

From these individual parts, slides could be made which would be transmitted while the revised script is read.

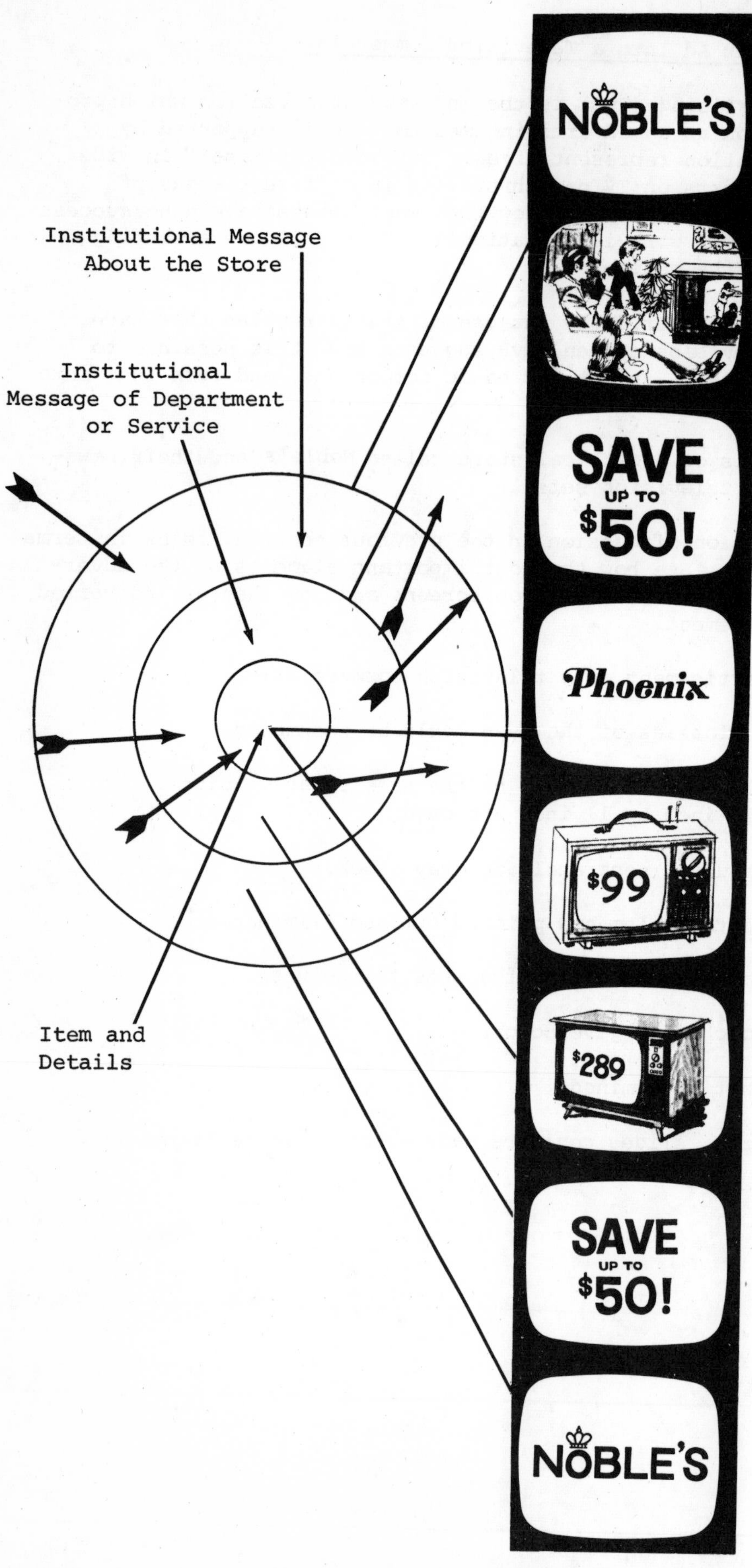

(1)
Noble's -- the store that saves you more.

(2)
Brings you savings from the wide selection in Noble's TV department.

(3)
With spectacular savings up to $50.

(4)
On famous-make Phoenix.

(5)
For example, this Phoenix portable for only $99.

(6)
And this TV console color set for only $289.

(7)
Save up to $50 on other Phoenix models. Sale priced at $199 and $149.

In the television department where the big brands are featured.

(8)
At Noble's -- Tomorrow -- the store that saves you more.

This Noble's commercial, an example of a simple television commercial, follows all the rules discussed in this chapter.

Going back to the explanation of the three basic parts of a television commercial:

The opening -- told about the store. What it stands for. Gave an institutional message.

The middle -- told about the merchandise. Featured the big benefit which was savings. The commercial stayed with the one central theme of savings right to the end. If the viewer got nothing else from the commercial, if she was not interested in a new TV set, she knew that it was Noble's, a place for savings. In this case, the big savings are in TV sets.

The closing -- which asked for the order. Told the viewers where to get the merchandise.

Going back to the section of the 5-S approach, all the points were covered of Simplicity, Sincerity, Sell-ability and Sold-where. There was no special Shock feature included except for the low price of the merchandise.

Adding Improvements to the Simplified Example

Many things could have added to the "stills" which made up the commercial to make it a better commercial.

Frames 1 and 2. They could have been combined. The store name could have been drawn over the artwork or supered over the art. The "sig" could have started as a dot and zoomed to fill the screen.

Frames 2 and 3. The picture of the family group could have been held while Frame 3, giving the price savings, could have been supered over it.

Frame 4. The brand name could have been given motion by zooming from a small signature to filling the screen.

Frames 5 and 6. When the copy on Frame 5 was concluded, the art in Frame 6 could have been wiped, thus eliminating the fast cut.

Frames 7 and 8. The savings copy could have been combined into one piece of art with the "sig." Or, Frame 7 could have come up first, and then dissolved into the store signature of Frame 8.

Other Improvements

Still working with the basic elements and not adding anything else, the camera could have moved on the art to give a feeling of motion. Zoom in and out on the merchandise. Zoom in and out on the family scene. Possibly start with the set in the family scene. Then pull back to show the family seated around the television set.

The addition of background music would add a touch of professionalism. Of course, the real improvement would have been to use the basic elements of the storyboard and do the commercials live in front of a camera, with a salesman-type talking about the sale and showing the merchandise.

Humor in Television Commercials

Good humor in advertising can be great. And, it can be disastrous. What is funny to one person might not be funny to somebody else. What is funny to one ethnic group can defeat the purpose of the advertising with another group.

A funny commercial, like a funny joke, is not funny after it has been told once or a few times at the most. The commercial becomes worn out in a short time. Perhaps this is why humorous commercials often win awards; the judges see it once while the commercial is new and fresh, not after it has played and replayed many times to the boredom of the viewer. Humorous commercials can be good once in a while. But beware of funny things that lay an egg.

Voice On -- Or Voice Over -- Which?

Should your actor on television do the talking? There are many answers to this question, but generally:

> Voice on -- If the actor can add impact to the commercial by speaking the lines, then the commercial should be delivered live by him. Actors who talk live on TV command a higher price than silent performers.
>
> Voice over -- If the actor adds little or nothing by doing the talking, then it is best to do voice over.

If the audio is voice over, the best procedure is to record the voice first before the television "shoot." Then, while the commercial is being shot, the audio is rolled so that the performers know what is being said while they are acting out the commercial. This means that the audio is completed, with music mixed with the voice, before going into the studio for shooting, and usually results in economies in the shooting because any corrections or changes were made ahead of time in the audio sound room when recorded, instead of using valuable studio facilities time.

Some cases where the advertiser might want to have the actor speak the lines would be:

- The commercial portrays a store buyer or other store executives. He should look and speak the part.
- The commercial portrays a salesman. He should look and speak the part.
- The commercial portrays a housewife. Again, she should look and sound as a believable housewife, not a Hollywood star.

CHECK LIST FOR EVERY TELEVISION SCRIPT

	✓
Have you intrigued your viewer with your opening?	_____
Is this benefit shown at the same time?	_____
If more than one item is shown (2 or 3) are they related to each other in merchandise or price?	_____
Have you avoided quick cuts, tricky camera angles which confuse the viewer?	_____
Are the words simple, easy to understand and believable?	_____
Are the performers believable for the part they portray?	_____
Do the words prove the value, quality, reliability?	_____
Is the one big benefit told and retold?	_____
And is it shown with a super?	_____
Are tight shots planned to show important features?	_____
Do the words match the character of the merchandise and the advertiser?	_____
Do the performers match the character of the merchandise and the advertiser?	_____

	✓
Have you avoided "written" words used conversational, first person, action words?	_____
And the video, too?	_____
Has a memorable scene been used and repeated?	_____
Does the audio and the video "say" the same thing at the same time?	_____
And shown in clearly?	_____
Have you read the copy aloud?	_____
Have you clocked it to check the length in seconds?	_____
Have you ordered cue cards or teleprompter?	_____
Have you a shock or surprise feature?	_____
Have you mentioned the company name frequently?	_____
Have you asked for action in the close?	_____
Has someone read it aloud to you?	_____
Does the script give video directions?	_____

Scripts and Storyboards

The following pages show various television commercial storyboards. The first commercial is shown starting with a script and its storyboard.

COLORFUL WHITE SALE

DRAPERIES AT $8.44

	VIDEO	AUDIO
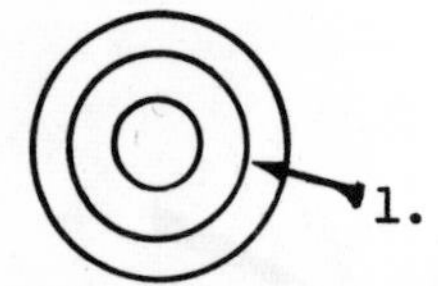	1. Split screen. Camera #1 picks up set. Camera #2 store sig. Music under soft tempo.	COLORFUL HOME FASHIONS...ALWAYS AT HOWARD'S. NOW -- AT COLORFUL SAVINGS.
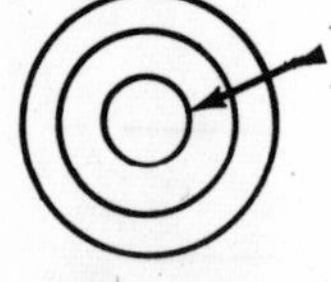	2. Drapery with table to give atmosphere.	SAVE NOW -- FAMOUS BRAND DRAPERIES.
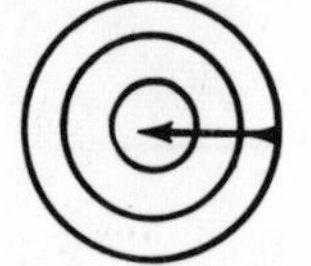	3. Cut to C.U. of drapery to show pattern and texture.	FOR EXAMPLE, THIS DAMASK WEAVE IN RICH, SOFT COLORS.
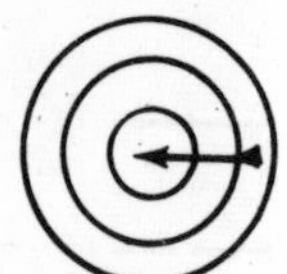	4. Cut to model beside window, holding back corner of drape, showing insulation.	SAVE DOLLARS -- SAVE WORK. THEY'RE MACHINE WASHABLE -- PERMANENT PRESS. FOAM INSULATION -- PROTECTION AGAINST HEAT, COLD AND NOISE.
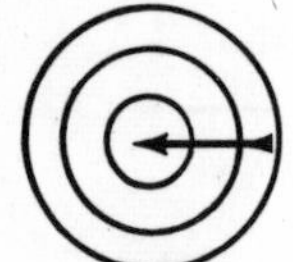	5. Cut to C.U. of drapery with table and lamp to establish feeling. Super price.	THE SINGLE WIDTH -- 63" LENGTH -- SALE PRICED AT $8.44. PRACTICALLY EVERY DRAPERY AT HOWARD'S AT COLORFUL SAVINGS.
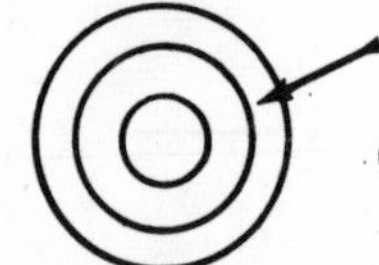	6. Cut to card with sig and reproduction of store's white sale poster. Music comes up at end.	DURING THE COLORFUL WHITE SALE -- NOW -- AT HOWARD'S.

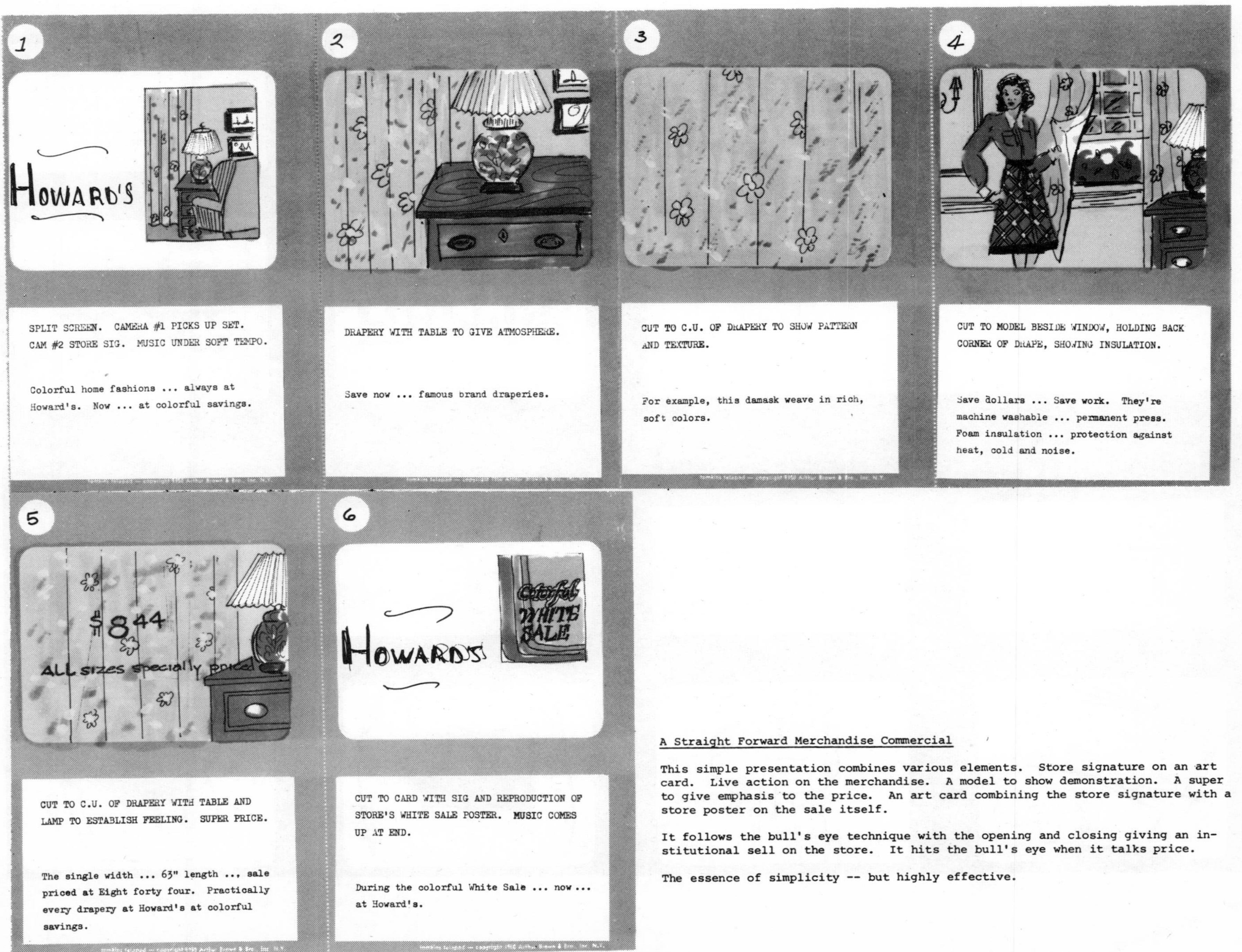

A Straight Forward Merchandise Commercial

This simple presentation combines various elements. Store signature on an art card. Live action on the merchandise. A model to show demonstration. A super to give emphasis to the price. An art card combining the store signature with a store poster on the sale itself.

It follows the bull's eye technique with the opening and closing giving an institutional sell on the store. It hits the bull's eye when it talks price.

The essence of simplicity -- but highly effective.

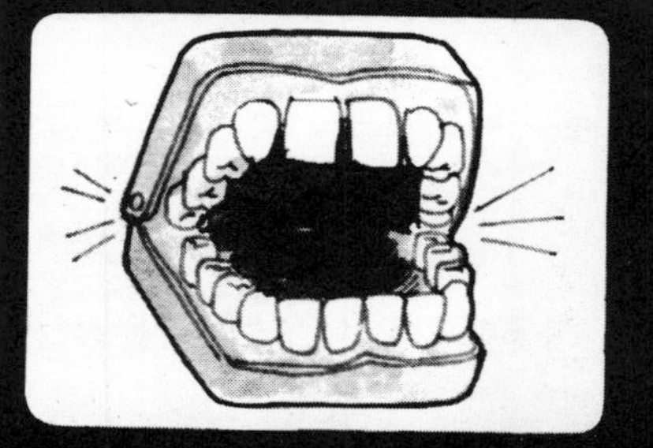

MED. SHOT OF A SET OF TOY CHATTERING TEETH. CAMERA BEGINS TO CLOSE IN.

Return anything to some stores and you're in for some heavy chatter.

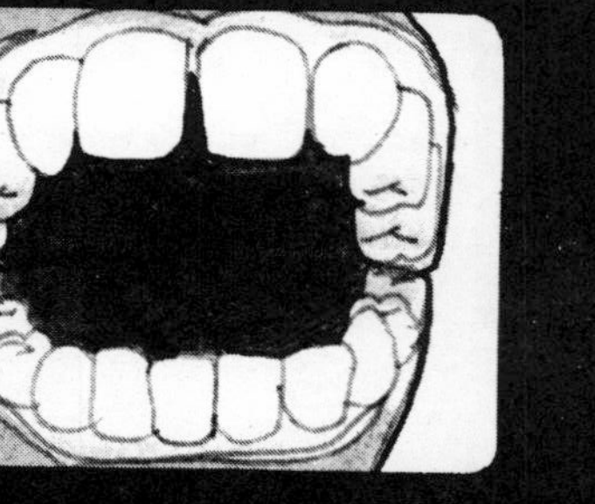

CONTINUE CLOSING IN TO CU TEETH BOUNCING UP AND DOWN.

Maybe even an argument.

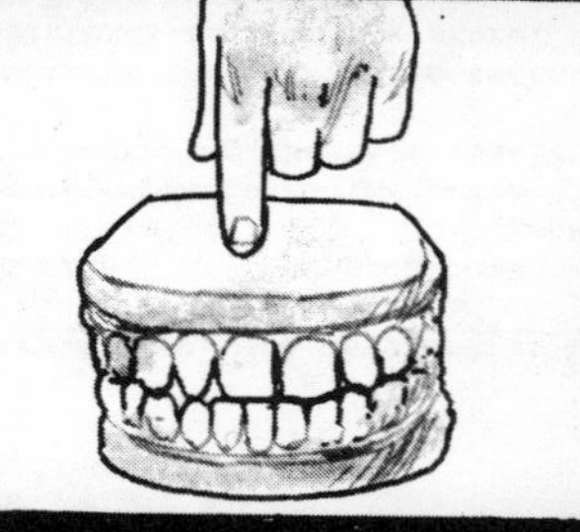

SAME SHOT. HAND COMES DOWN AND SHUTS MOUTH WITH ONE FINGER.

Not at Harris'. You have a problem -- we listen. And 9 times out of 10 -- give you exactly what you want.

A HANDFUL OF MONEY IS PLACED NEXT TO TEETH.

Your money back. Or a credit. It's just one more reason to shop Harris'.

CUT TO STILL PHOTOGRAPHS OF THE SALE MERCH. SUPER: 20% OFF FASHION SLACKS 9.98 AND UP.

Like our sale of all men's fashion slacks, 9.98 and up. Through Saturday

Harris'
We're not the biggest, so we try to be the best.

HARRIS' LOGO AND TITLE: WE'RE NOT THE BIGGEST, SO WE TRY TO BE THE BEST.

At Harris'. We're not the biggest, so we try to be the best.

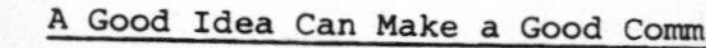

A Good Idea Can Make a Good Commercial

The props for this commercial were the chattering teeth available in any joke store and a stack of money.

The commercial follows the bull's eye technique. The first four frames are the opening of the bull's eye: a combination of a surprise and institutional opening. Then we move to the center and hit the bull's eye and sell the merchandise. The commercial ends with the institutional closing.

The fifth frame is the only one which sells merchandise. It could be expanded to show more merchandise holding the "20% off" super all the time. Or frame #5 could have been followed by other men's merchandise such as shirts or shoes.

The opening and closing can be used as a sandwich commercial for any number of items at different times for the "Harris" store. The opening four frames and the last one could be produced on tape or film. The item section could be added each time there was a merchandise change. This idea could be adapted in many ways using things from a joke store as props.

WIDE SHOT SHOWING BIG TOY MECHANICAL FISH CHASING SMALL FISH

When you're not the biggest store in town you have to do more to stay ahead.

CLOSE UP OF SMALLER FISH

Like give better service. And better values. If you don't ...

3/4 SHOT OF BOTH FISHES.

the next guy will. At Harris'

3/4 SHOT WITH LITTLE FISH IN FOREGROUND TRYING TO STAY AHEAD OF BIG FISH

we've been doing more for nearly sixty years. And we're not about to quit now.

20% off sheets and spreads

CUT TO STILL PHOTO OF MERCH.
SUPER: 20% OFF ALL SHEETS AND SPREADS.

This week's sales are 20% off every sheet and bedspread in the store. Pretty darn good!

Harris'
We're not the biggest, so we try to be the best.

HARRIS' LOGO AND TITLE:
WE'RE NOT THE BIGGEST, SO WE TRY TO BE THE BEST.

Even for Harris'. We're not the biggest, so we try to be the best.

Another Good Idea for an Inexpensive Commercial

The props for this commercial are toy fish which are available in typical toy stores. This commercial also follows the bull's eye technique. The first four frames combine a surprise treatment together with an institutional message. This is the outer rings of the bull's eye. The next frame is where you "hit the bull's eye" and make the sale. The last frame is the outer ring of the commercial where the "Harris" store is given an institutional closing.

This is also a sandwich type commercial. The opening and the closing could be created on tape or film. The item part could be changed to meet merchandise requirements.

As in the chattering teeth commercial, the merchandise part of this one could be extended. The camera could go from a long shot of the merchandise to a close-up as shown. Or, other related items could be shown. A group of sheets and cases, then a grouping of bedspreads. Instead of the 20% off theme, prices could have been shown separately. This idea could be adapted in so many ways using toys as props.

#1

MEDIUM SHOT OF SCHOOL DIRECTIONS LOGO. FADE UP FROM BLACK AS CAMERA PANS RIGHT, REVEALING WORDS AS IT GOES.

Follow the school directions.

#2

CONTINUE PAN RIGHT TO END OF ARROW.

School directions.

#3

"SHOP" SIGN IS REVEALED AT THE END OF THE ARROW.

Shop

#4

ZOOM IN TO MEDIUM CLOSE-UP OF "SHOP" SIGN. HOLD FOR YOUR STORE SUPER.

Store Name for

#5

LOSE "SHOP" SIGN, ZOOM IN TO KEY OF MERCHANDISE UNTIL IT FILLS THE SCREEN.

Pants by Mann of 50% polyester/50% cotton.

#6

MERCHANDISE

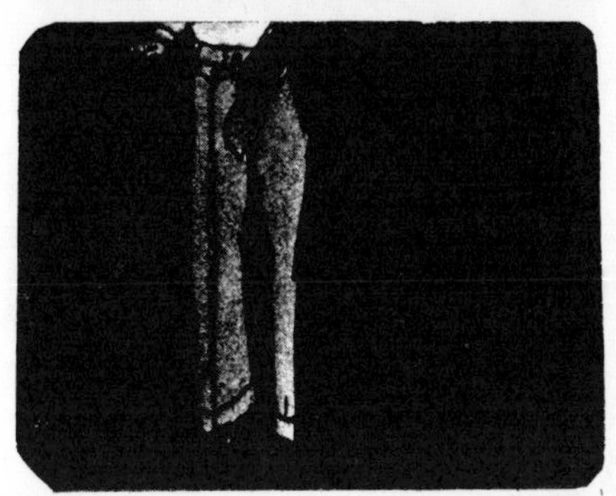

Sizes eight to twenty.

#7

MERCHANDISE

Baggies, just six dollars.

#8

MERCHANDISE

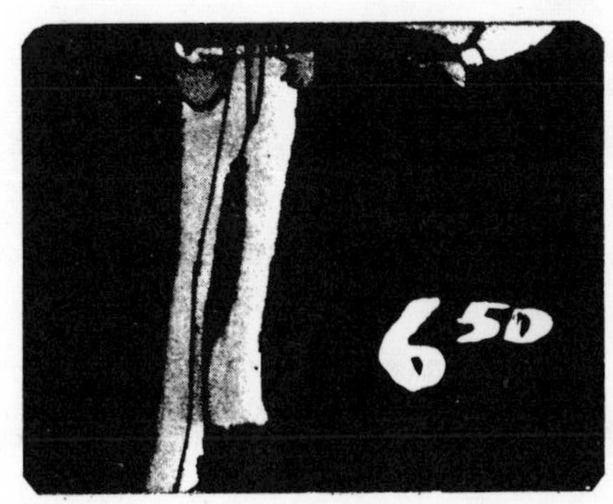

His favorite, flared jeans, just six fifty.

#9

MERCHANDISE

Denim striped cotton knit shirts just four fifty.

#10

LOSE MERCHANDISE, ZOOM OUT TO MEDIUM CLOSE-UP OF "SHOP" SIGN, AS IN #4.

#11

PAN RIGHT TO ARROW READING "SCHOOL DIRECTIONS."

Follow the school directions.

#12

PAN TO END OF ARROW, SUPER STORE NAME.

Shop Store Name, your back-to-school headquarters, of course.

DECISION #10

Producing Your Television Commercial

Television production can vary all the way from a set of inexpensive slides to on-location shooting in a foreign country with dozens of actors, original music, fancy sets, involving lavish expenditures of every kind. Actually, the key to production is not lots of money, but a creative idea.

Most commercial shooting is done in studios under controlled conditions. However, there are big advantages of doing commercials on location. This depends on the purpose of the commercial and the merchandise it portrays. Example: Swim suits and sportswear are obviously best on location. In shooting on location there is the advantage of nature's own lighting.

In this production section, these points ae considered:

- The tools of television -- stills, tape, film.
- Tape or film.
- Importance of lighting.
- Camera and studio effects.
- The sandwich short cut.
- Selecting music and props.
- Selecting talent.

The Tools of Television

Stills continue to be used for many commercials. Still art includes slides, art cards and transparencies. In using stills other than slides, some standards should be considered:

- Type -- strong. No serifs. Clear. Well-spaced.
- Working area -- 11 by 14 inches.
- Bleed area -- 9-1/2 by 13-1/2 inches.
- Artwork area -- 6-3/4 by 9 inches.

Slides are made by shooting 35mm film. Developed. Mounted. For just a slide commercial, they are usually put in a fixed position in a slide projection chain and transmitted while the copy is read.

However, for better production, a transfer can be made to tape including the audio. Then transmission only depends on rolling the tape.

A GUIDE FOR PRODUCING SLIDES FOR TELEVISION

Length	10 Seconds	30 Seconds
Words	20 to 25	60-65
Slides	Two to Three	Six to Eight
Logo	Final Slide	At Opening and Closing

Art Cards. Different kinds of art are used. For actual showing of merchandise, art can be drawings, photographs, type and combinations of them. Camera can move on the type to give a feel of motion with zooms, panning and tilting.

Transparencies (ektachromes) are often used. Adding light to the background gives them a more realistic feeling to the viewer. A feeling of depth. Shadows seem to disappear. Camera movement can be used on any stills except for slides.

Supers are made on art cards. When the background scene is light, the supers are black on light gray stock. When the background is dark, the supers are white on black art board.

Store signature is generally on an art card. If the regular "sig" is fussy, it should be simplified.

CHECK LIST FOR TELEVISION ARTWORK FOR EACH COMMERCIAL

Art to be prepared by ______________________________

To be created as: __ Slide __ Photo __ Drawn __ Transparencies

To be okayed by ______________________________

Deadline for studio ______________________________

Check to see if art looks good both in color and b/w ______________

Check legibility of type ______________________________

Check for contrast in color and gray tones ______________________

Should You Use Tape or Film?

Sometimes the choice of tape versus film depends upon what is available in your own market. If the television station is the best source of production, chances are that you would use tape, except for on-location shooting. On the other hand, if there is a good, even though small, production film company that has acquired a knowledge of local production, film might be the best way to go.

FILM SHOOTING

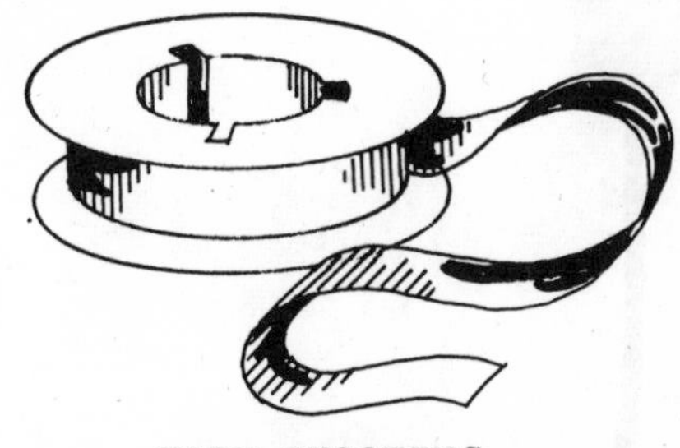

TAPE SHOOTING

FILM SHOOTING	TAPE SHOOTING
More portable cameras.	Although new backpack cameras almost measure up to film mobility.
Very flexible for location.	Although mobile tape units exist.
Capable of wide range of optical effects.	Although tape has its share and created automatically and immediately.
Simpler shipping problems.	Tape is bulky.
Cheaper to make dubs.	Much faster for editing, processing.
Color system -- yellow, green, blue primaries. TV must transfer these colors in transmission.	Color system -- blue, green, red primaries. No translation needed in transmission.

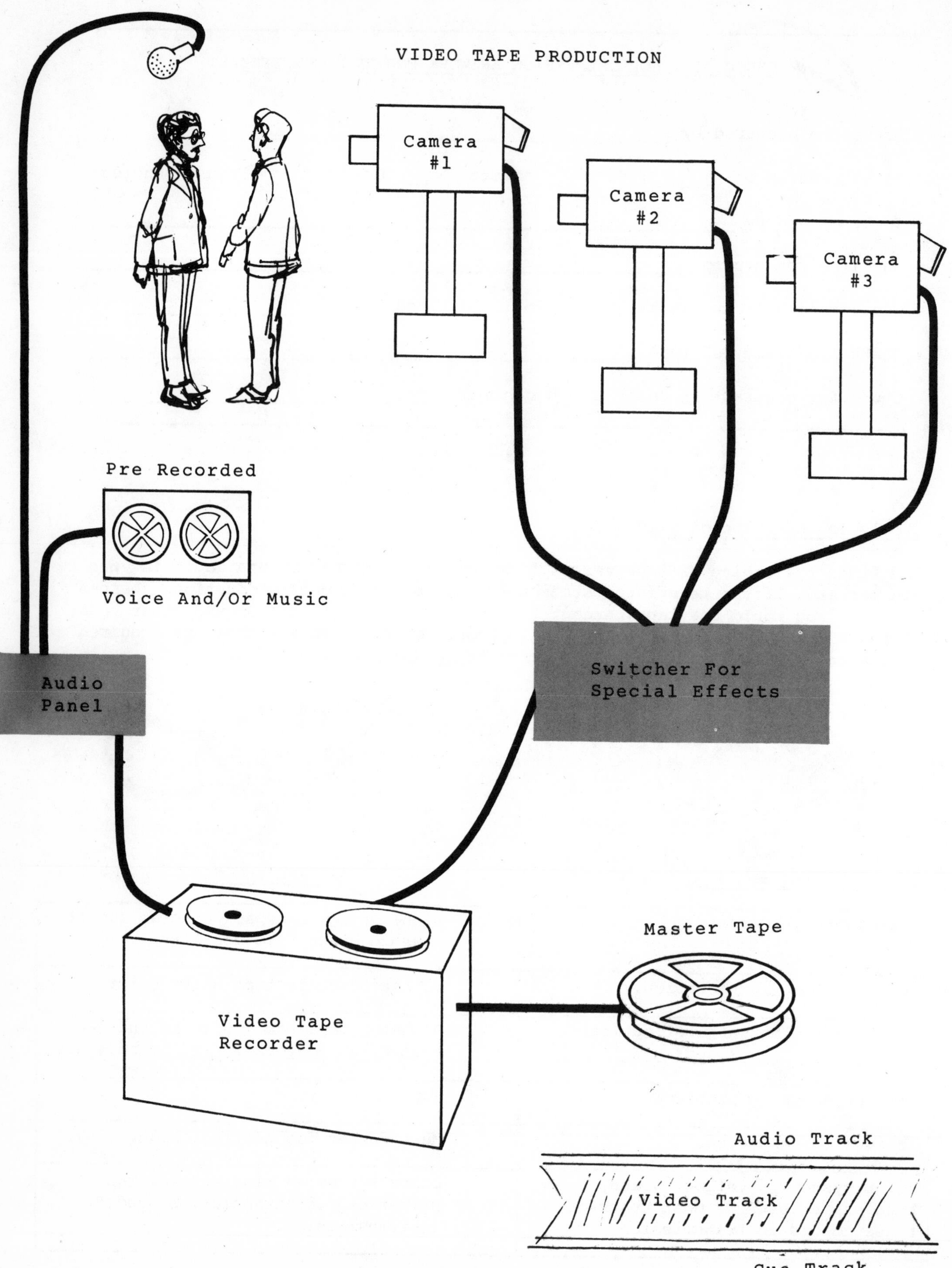
VIDEO TAPE PRODUCTION
Camera #1
Camera #2
Camera #3
Pre Recorded
Voice And/Or Music
Audio Panel
Switcher For Special Effects
Master Tape
Video Tape Recorder
Audio Track
Video Track
Cue Track

FILM PRODUCTION

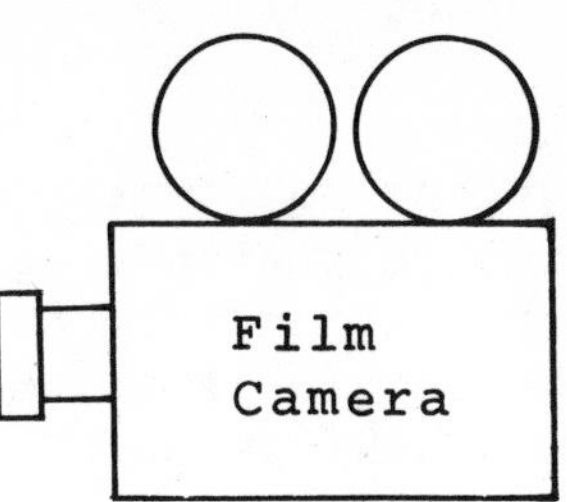

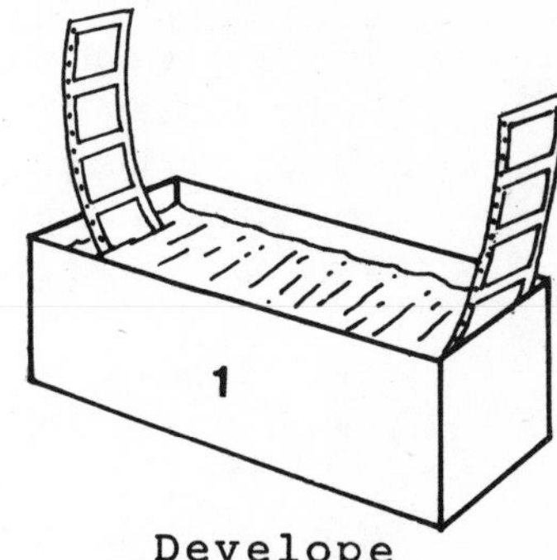

Develope

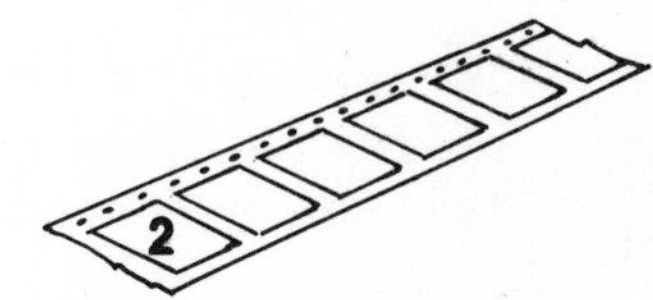

Select The Takes Wanted

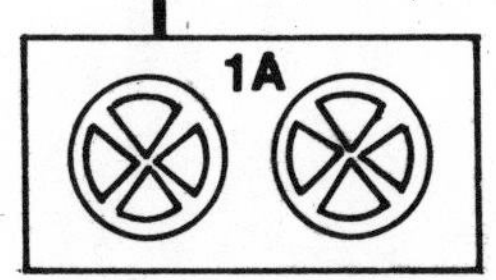

Tape Recorder
Voice and/or Music

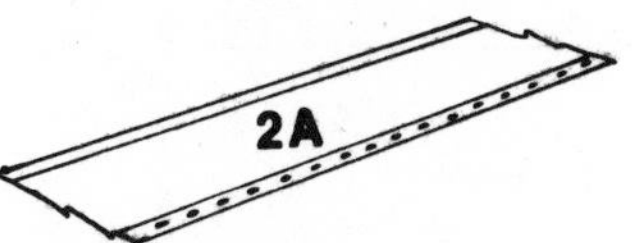

Transfer Tape
Sound To Film

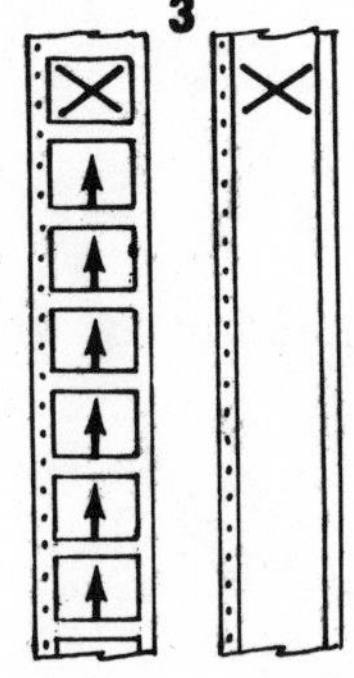

Match sound
with picture on
moviola or steenbeck
and indicate for editing

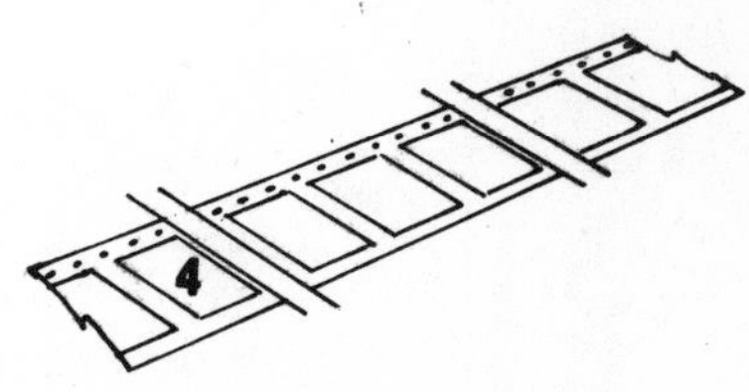

Select and edit
sound and pictures

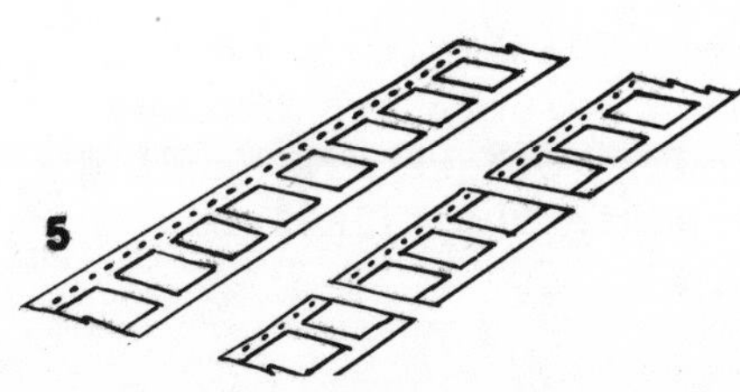

Match up and edit for workprint.
If magnetic track was used
previously, convert
to optical track.

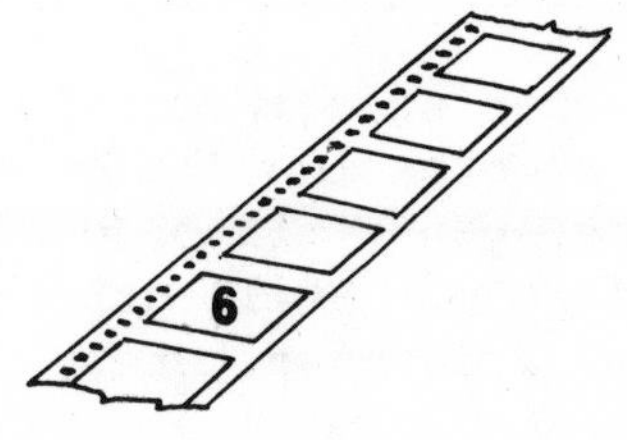

Make answer print
determine any changes
for color or timing

After adjustments
make final
release prints

Overhead flat lighting gives even lighting for the overall scene. TV cameras, using flat lighting only, pick up wrinkles and folds in garments.

Key lighting and cross lighting modify the effect. Example: Key lights give a subject dominance within a scene. Assist in setting mood. Cross lighting can be strong from the side to show fur and fabric texture. Cross lighting is also used for dramatic impact, dimension and perspective. Fashions on TV should combine the effective use of flat, key and cross lighting.

Full lighting will eliminate shadows and gives a flat level of light. The full lights should come from the sides.

Back lighting brightens the merchandise. It tends to give a separation from the background of the set.

To create the brightness or the mood required, think of the tricks of the window display man in a store and how he manipulates the lights to get the best effect for the merchandise.

✓ CHECK LIST FOR LIGHTING FOR EACH COMMERCIAL

What mood is wanted for the commercial? High fashion ______________

Hard sell ______________ Soft sell ______________

Are special color lighting needed to create a mood? ______________

Warm effect ______________ Cool effect ______________

Does the studio have all the lighting equipment needed for these effects -- or will it have to rent it? ______________

Effects That Can Be Created in Shooting

The effects obtained in shooting commercials are endless. The limit is only in the imagination of the team creating the commercial. Some effects are through the camera man. The zooms, pans, tilts, dollies in or out. Others are handled in the video tape control room. Still others are obtained in the film labs. Some of the most used ones are shown here.

Much is said for better color control in tape and also for the speed from shooting to final. Sometimes it becomes practical to shoot in film, edit in tape and create the final commercial in tape. Tape transfers to film are relatively simple and inexpensive. Tape transfers to film, until recently, have been generally unsatisfactory for air quality, but there have been major improvements in quality and a reduction in price. There are still some purists who shoot commercials on 35mm although 16mm is the standard for television.

Shooting Charges for Tape and Film

The typical production studio bids on the TV commercial job on the basis of doing the total job whether the house is a film or a tape house. Television studios, however, often have a charge based on the hour or the day. This includes the complete facilities of the studio, studio assistants and camera people, prop assistants, a director, and use of the control room. Prices vary considerably from market to market.

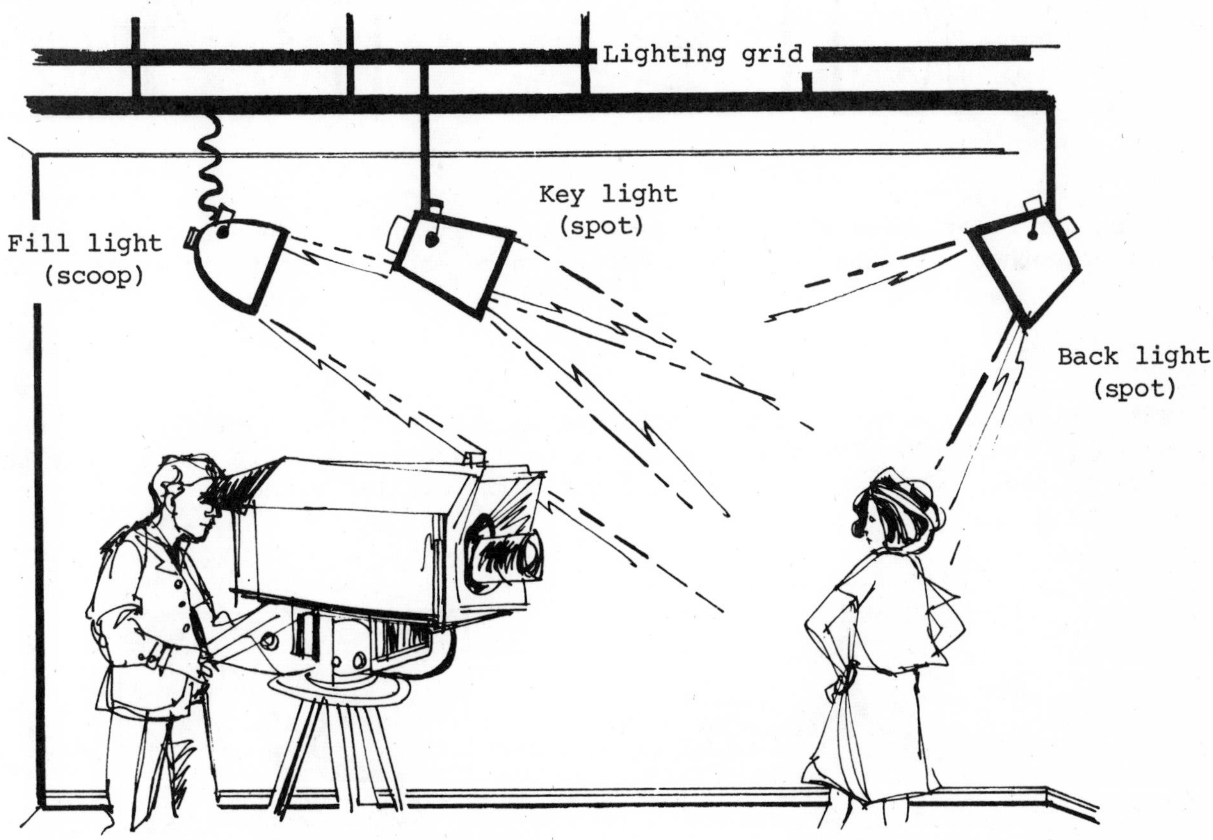

Importance of Lighting for Good Commercials

Lighting is used, not just to show the merchandise, but to show and give emphasis to the features of the merchandise.

Lighting is used to create a mood. In the early days of television, flat flood lighting was used. It gave a single brightness, no contrast lighting, no shadows. It allowed for no lighting emphasis where the artist wanted special effects. In recent years, the studio engineer gave way to the artist in lighting selection. Lighting became almost as important as the camera itself.

A studio camera is a movable object. It can move to and away from the objects. With its zoom lens it can go from a pin point to fill an area and then the reverse. It can tilt to show different portions of the scene. It can move from side to side panning the set or item. The shots used can go from extreme close ups (ECU) to a long shot (LS) depending on what you want to feature.

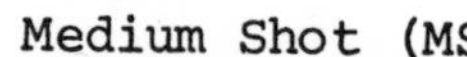

Close Up (CU)

Medium Shot (MS)

Long Shot (LS)

Wipes

Wipes can be achieved with one object originally shown. Then it is "wiped" away while another replaces it. These wipes can be made horizontally or vertically.

Split Screen

This is another effect which is obtained automatically with tape. One camera picks up one shot. The other camera picks up another scene. The two are shown side by side. Some of the other automatically created effects:

Venetian Blind

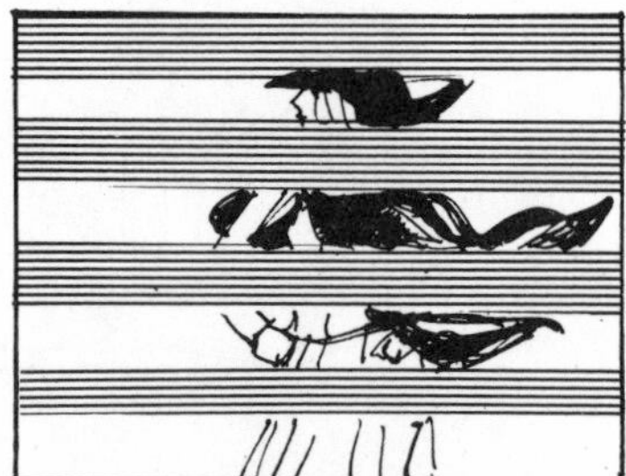

Saw Tooth

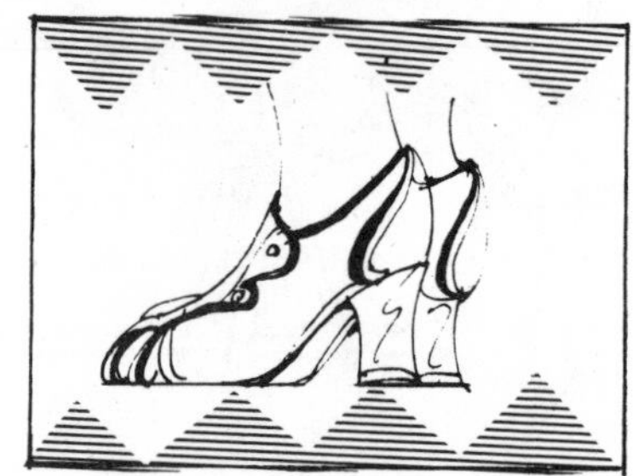

Diamond

Port Hole

Cuts

Using more than one camera, the video tape control room can make automatic cuts going from one piece of merchandise or situation to another. Of course, with film, this can be accomplished in the lab. Cuts suggest fast movement and are adaptable for hard sell or to create a fast movement or an elapse of time.

From This

Cut to This

Dissolve

To obtain a smoother transition, however, the dissolve is used. This is a gradual movement where the first and second scene mix into a blur followed by the first one eliminated and the second one sharp. A switcher in the control room of the video tape studio does this. The pace is much slower than the cut and is adaptable for fashions. In film, this effect can be obtained in the labs.

Chromakey, a Mixing of Camera Shots to Achieve Effects

Chromakey is used to achieve a wide variety of situations. Almost everyone has seen detergent commercials with the enormous box and the actor looking small alongside it or on top of it. There have been auto commercials where miniature mechanics are seen all over the car.

An outdoor scene of the beach, a country club golf course, a local landmark can have the models right there in the picture with this effect. The subjects are shot against a colored background, usually green or blue. The other camera picks up the picture of the location. Merchandise can be made to appear in any size in almost any part of the scene.

Chromakey blue background

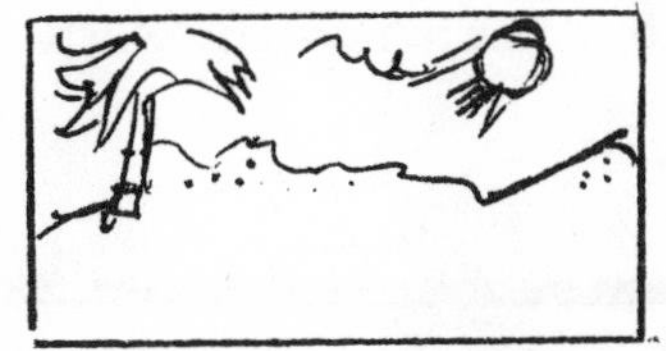

Chromakey to this

Both Rear Screen and Front Screen Projection

Slides, photos, artwork, and even motion pictures can be projected from either the rear or from the front to establish a "set" inexpensively. The "set" can include the interior of the store to give the impression that the salesman is right in that department. The "set" can place your models on Fifth Avenue, Michigan Avenue, Paris, Hollywood, anywhere. The art or photograph is projected with one device while the camera takes the subject.

New Techniques

The father and grandfather of television techniques were the theater, the movies and radio. Over the years these forms of art had constant changes. TV techniques are continually changing and improving. As part of the continuous education of a TV writer, he should visit stations and studios regularly.

The Sandwich Commercial -- A Production Short Cut

The sandwich commercial (sometimes called a donut commercial) is a short cut used by many local advertisers. With the sandwich you eat, the filling is between the two slices of bread. Translated into television, a television commercial has an opening and closing. These are the two slices of bread. The middle of the commercial is the filling.

Sandwich commercials permit you to spend time, effort and money on both ends to create a good message. Then, on a day-to-day basis, you add the merchandise you want to advertise in the middle, the sandwich filling. Since the opening and closing are expertly created and produced, it is possible, if needed, to have the middle produced inexpensively. The openings and closings create quality effect for the total commercial.

The stores that create sandwich commercials usually prepare several of them. This is much the way you vary the bread you use in the sandwiches you eat. Today with white bread. Tomorrow with rye or whole wheat. Changes can be for sandwich commercials for:

- Men's departments.
- Children's departments.
- Fashion departments.
- Home furnishings departments.
- Institutional or service selling.
- And sandwich commercials for major events throughout the year: January White Sale, February Furniture Sale, right through the year to Christmas commercials.

In use, the sandwich can have different variations:

......7 Seconds......	............20 Seconds..............	.......3 Seconds......
SANDWICH INTRO	SPECIALLY PRODUCED MIDDLE	SANDWICH CLOSE
Sells the store or a sale	Specific merchandise in the event	Institutional about store, the "sig" and address

But there are times that you are only hungry for an open sandwich with one slice of bread:

...................27 Seconds................	..........3 Seconds.............
SPECIALLY PRODUCED OPENING WITH MERCHANDISE/EVENT, ETC.	SANDWICH CLOSE
Come on directly with the merchandise or the sale event plus the merchandise	Use the ready-made sandwich ending

The intro and close could be combined to create a 10 second commercial to sell the event, sale or service...

..................10 Seconds..................	
SANDWICH INTRO	SANDWICH CLOSE

...in which case the audio has to be re-recorded for the full ten seconds of even flow.

.....3 Seconds...........	....................27 Seconds...............
SANDWICH CLOSING USED AS THE INTRO	SPECIALLY PRODUCED MIDDLE AND ENDING
Use the ready-made segment meant as the close	Go into the merchandise or the event/ merchandise and conclude with the store signature and where to buy

Special Advantages of the Sandwich Commercial

There are many big advantages in using the sandwich idea for the typical local advertiser:

- Provides an institutional (or sale) repetitive message at the intro and closing.
- The repetition of this message sells the store or event or service through frequent use.
- Provides an easy base for item commercials. Only the middle must be created at last minute timing.
- Although the original production costs for the sandwich commercials are high, their frequent use drastically lowers the cost each time used.
- Gives a "look" equal to the best on the air because of excellent openings and closings.
- The middle part can be created in the same way as regular production -- film, tape, stills.
- Different sandwich commercials for different events and different departments can be created well in advance -- all with the same family "look."
- And furnishes an excellent vehicle for co-op advertising. Each co-op ad is held together with one "look" through the use of the same (or similar) openings and closings.

Selecting Your Music for Television

The writers of the big Broadway musicals know the value of a theme song. They strive for a central theme expressed in song and music and play it early in the show. They will repeat it during the various scenes. When you leave the theater the orchestra plays it as a finale. If the audience is not humming the tune as it leaves, musical comedy writers know they have a problem.

In the same way, a store can have its own musical identification, its own sound. A jingle can set a theme for the store and be a great asset. The words and music could become as popular in the markets used as the top hit record of the period. There are many examples of how television theme jingles have done this, such as Pepsi, Coke, Avon, National Airlines and Robert Hall.

Jingle writing is an art in itself. No amateur should attempt it. Leave it to the various companies, small and large, which can create a jingle to meet the specific need. The jingle house should be told:

- The quality, flavor, standards of the store.
- The general directions of the store.
- Slogans used by the store and those preferred.
- The kinds of merchandise which will be promoted using the jingle.
- The period of time the jingle will be used.

While one basic song, words and music would be created, it is wise to orchestrate it in different ways to match the many kinds of merchandise to be used. One version for fine fashions. One for men's merchandise. One for teens. One for home furnishings, and so on. Music can be helpful in almost any commercial to set a mood. The fast tempos for hard selling. The quiet music for fashion selling.

Availability of Library Music

Instead of resorting to original music, most local advertisers obtain their music from music libraries. These are libraries which have music of every possible description and sell the use of it. Stations often have their own music libraries which they allow their advertisers to use for a modest fee.

Warning! Always check to see if you have complete clearance to use library music. Any reputable house will furnish you with the proof.

In recording sound for the audio voice, the usual procedure is to play the music while recording the audio voice and mixing it at the same time.

Handling music in a commercial is an art itself. By experimentation it is possible, even for the beginner, to see where to come up strong, where to go under, where to come back strong and when to fade out.

Choosing Your Talent for the Commercial

Casting is important. The right person for the commercial is essential to make the commercial look right and to make the commercial believable to the viewer. A Hollywood star, for example, would be out of place in ordinary dress or demonstrating a washing machine. A truck driver would be unbelievable in a tailored suit. When in doubt, always underplay.

If the model does the audio in front of cameras, the voice must be just right for the part. Generic voices are always better than ones with special or ethnic accents. In some cases, the performer must memorize the audio. In other cases cue cards or teleprompter devices are used. Be sure the performer can read without glasses if he has to read from cues.

Models for television should move slowly, never abruptly. A comfortable appearance on camera makes for believability. Models should be shown in clothes which give the viewer the feeling that they are the models' own clothes and they should take normal strides on camera, not fashion show poses.

Almost every city has a model talent pool or agency. In some cities there are "little theater" groups which are often suitable for television. One thing to consider is that the "pro," although more expensive than the amateur, is often the cheapest in the end. They know what to do and when to do it with a minimum of instruction.

When using children, it is a good precaution to check the local labor laws.

✓ CHECK LIST FOR TALENT SELECTION

Who will select the models and do the casting?____________________

Who will handle hairdressing?____________________

Before the shooting__________ During the shooting__________

Is pressing required?____________ Who will do?____________

Where____________ When____________

Is a wardrobe girl needed in the studio?____________

Are the studio dressing rooms adequate?____________

If the models are to wear their own clothes, have they been told what kind to wear?____________

Are release forms ready for signing?____________

Choosing Your Sets and Props

When television commercials began, there was a tendency to go lavish with sets and the props. Gradually, TV producers learned from the retail display man. He knows that a display must be kept simple so the merchandise holds the chief emphasis. For example:

- The set must reflect the merchandise itself and not be too lavish when the merchandise is simple or inexpensive.
- Avoid complete room sets -- just use a corner or a segment.
- Watch the colors to be sure of contrast. Take a Poloroid in black and white to see how the set will come over in b/w television sets.
- Avoid splashy, fussy, thin stripes or other confusing backgrounds.
- The big key is the word K.I.S.S. -- Keep It Simple, Stupid,

CHECK LIST FOR PROPS AND SETS

Have you planned them with utmost simplicity?______________________

Who will design the props and the sets?__________________________

Have they been given full instructions?__________________________

Have they seen the script?_______________

To where do the props get delivered?____________________________

When____________________ By Whom___________________

Is insurance needed?____________________

Have arrangements been made to break down the set and return props?_______

Keeping a Control on Production Costs

Unless every step of television production is watched, it is possible to have the costs get out of control. The best way is to create estimates, as accurate as possible on each item of each planned commercial. These items include the studio facilities, the cost of the talent, music, graphics, all of the props. Naturally, there are other costs, such as trucking props to a studio or to "on location" sites. If additional tapes or film dubs are to be made, it is necessary to estimate them as well. Transportation charges for talent, crew and supervisors are often overlooked in planning a budget. They should be included.

If the "shoot" is with film, and on location, there are no studio facility charges. In this case, the camera crew charges can be estimated and shown in the "studio facility" part of the form or under "additional."

When a commercial is part of a group being shot at the same time, and at a combination price, a separate estimate sheet should be created just the same. Then the entire job, all the commercials, can be shown at the bottom of the form.

As quickly as the bills are received, the "actual" column should be completed and the cost control concluded. A systematic record-keeping procedure of this kind has the advantage of holding a better control on every expenditure which goes into the total cost. In addition, it provides a reference point for planning future commercials. This is especially true when planning similar events or merchandise.

Production Charges

Merchandise or Service____________ Event____________

On Air________ Need By________ Station(s)________

Code #________ Studio________ Shoot Date________

Responsibility at Advertiser____________________

Responsibility at Studio____________________

Is this commercial one of several in the same shoot? If so, how many in total shoot?________ Do a separate cost sheet for each and give total for all in last line.

	Date________ ESTIMATE	Date________ ACTUAL
Studio Facility $ ________	$________	$________
Hours______ at $______ Each	________	________
Studio Facility $________	________	________
Hours______ at $______ Each	________	________
Talent________	________	________
Use______ Rate $______	________	________
Talent________	________	________
Use______ Rate $______	________	________
Talent________	________	________
Use______ Rate $______	________	________
Music $________	________	________
Props $________	________	________
Graphics $________	________	________
Including art cards		
Dubs______ at $______	________	________
Shipping $________	________	________
Talent Transportation $______	________	________
Talent Expenses $______	________	________
Other Transportation $______	________	________
Additional ________	________	________
Additional ________	________	________
Additional ________	________	________
Sub Total	$________	$________
Tax	________	________
TOTAL	$________	$________
If this is part of a group of commercials: — Sub Total	________	________
Tax	________	________
TOTAL	$________	$________
TOTAL NUMBER____		

Television Commercial Production Schedule

A production schedule varies according to the quantity of commercials a store produces.

For television geared to events, it is possible to do planning far in advance. For example, if you know you will have a January White Sale, the basic look of the commercials can be planned many months ahead. In fact, the White Sale can be planned with changes so the commercial can be used for the May and August White Sales. Of course, the flavor of the commercial should tie in with the advertising done in newspapers, direct mail, posters and windows. The actual items used in the commercials would be selected closer to the time of the promotion as in newspaper advertising.

Similarly, all basic planning on commercials for events could be determined months ahead. It is realistic to create rough storyboards for all event commercials for a half-year at a time. The actual commercial can be created subsequently.

For a store with a regular, steady television campaign based on items, which may be part of an event or not, this time table is appropriate in those cases where video tape is used. For film production, a somewhat longer lead time is needed for editing, developing and printing films. Implementation of this time table should be within the framework of the store's own sales promotion planning. Item planning for television should originate no later than two weeks before presenting on-air when video tape is used.

Time Table	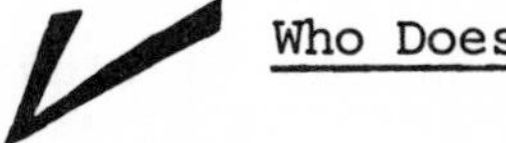 Who Does What?
First Monday	
Planning meeting to select specific merchandise to be on-air in the week starting two Sundays away. Followed by meeting with copy people, graphics, display and advertising agency if used.	______________
First Tuesday	
Discussions of copy approach with copy, graphics and display people and agency, if used.	______________
Examination of TV scrap book and files to see if any old commercials or ideas can be used or adapted.	______________

Who Does What?

First Wednesday and Thursday

Copy written, storyboards drawn. Other graphics including supers. OK's obtained from department heads, buyers. ____________

Display department notified for props. Sign writer notified for departmental signs. ____________

Display department notified for possible window or interior store display tie-ins. ____________

First Friday

Models selected. ____________

Music selected. ____________

Audio recorded, if voice over. ____________

Second Monday

Run through rehearsal. Possibly in the store to avoid studio time charges. Any corrections made during rehearsal. ____________

Second Tuesday

Production in studio. ____________

Second Wednesday

Any additional production, if needed. Meanwhile, all editing completed. Supers included. ____________

All buyers, departments informed. ____________

Second Thursday

Commercials available for department head to see. Commercials sent to stations. ____________

All department signs completed. ____________

Window and interior display tie-ins completed, ready to install at appropriate time of selling. ____________

Naturally a longer production period is better. But for an advertiser in a single market using TV on a year-round basis, this time table is realistic.

DECISION #11

Buying Your Television Time

In newspapers, the national advertiser buys space based on an "open rate" for the specific paper. There are extra charges for:

- Premium positions.
- Special pages.
- Magazine sections.
- Color pages.
- And other features.

Some papers offer "bulk" rates to national advertisers who contract for a certain number of pages or lines. Space charges are commissionable to advertising agencies.

In newspapers, the local advertiser is usually offered a lower rate than the national advertiser. With the local advertiser, these rates are usually not commissionable to an advertising agency. Rates have many variations such as:

- "Bulk" contracts for a pre-determined amount of space within a year.

- Frequency plans. The advertiser gets a reduction if he runs a certain amount of space each week, or runs advertising a certain number of times a week.

- A higher rate for color pages.

- A higher rate for the magazine section.

- A different rate for various sections of the paper.

- A higher rate for premium positions.

Typesetting is usually free. When corrections are made above a certain amount, papers generally have a charge. Some newspapers make free cuts or give the advertiser an allowance for cuts based on the amount of space used.

With newspapers, rates vary from paper to paper. In competitive markets, the circulation of rival papers determine their space rates. For example, in New York there are two tabloids, the Post and the News. The News, with its much larger circulation, charges considerably more for space. After all, advertising buys people, or the opportunity to reach people.

In television, pricing of commercial time follows the same basic standards used in newspapers. Time costs are based on audiences reached. Earlier in this book, viewing charts were shown of households reached throughout the day. The pricing charts show how rates increase when audiences increase and drop late at night when the audience drops.

In the most elementary TV rate card, the rates vary according to the time of day. These dayparts are broken down from three to seven different parts depending upon the station. Each part receives a letter designation. Each is based on the size of the audience.

Although these dayparts were explained earlier, they are shown here for the typical station. Keep in mind that, similar to the way pricing methods vary from newspaper to newspaper, these TV daypart designations also vary from station to station. It is a good idea to ask the stations for their exact breakdowns of dayparts.

Class "AA"	7:30 PM - 11:00 PM Monday through Saturday 6:00 PM - 11:00 PM Sunday
Class "A"	7:00 PM - 7:30 PM Monday through Saturday 5:00 PM - 6:00 PM Sunday
Class "B"	6:00 PM - 7:00 PM Monday through Saturday
Class "C"	9:00 AM - 6:00 PM Monday through Saturday 9:00 AM - 5:00 PM Sunday 11:00 AM - 12:00 Midnight, daily
Class "D"	Sign-on to 9:00 AM, daily 12:00 Midnight to sign-off, daily

In the simplest rate structure, for each of the dayparts, prices of the commercials would be indicated with the designated letter. However, all stations have some shows which reach larger audiences than other shows in the same dayparts. The typical station, therefore, adjusts the rate higher for those shows where the audience is larger and lower where the audience is smaller.

Rates Are Based on Lengths of Commercials

When you buy newspaper space, you buy by the line or inch or page. When you buy television, the length of the commercial is also a determining factor. In television, there are these lengths:

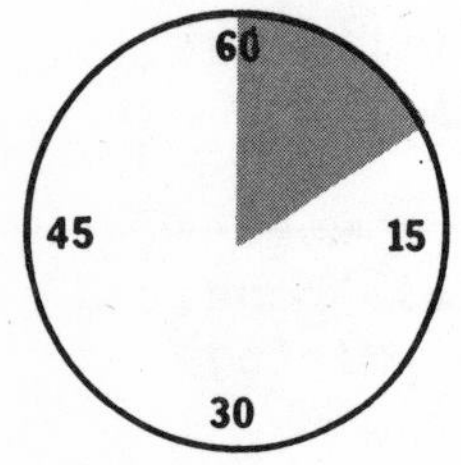

- 10 Seconds. These can be full ten seconds or can be slightly shorter as an ID (station identification) commercial where the station uses a few seconds to show their call letters.

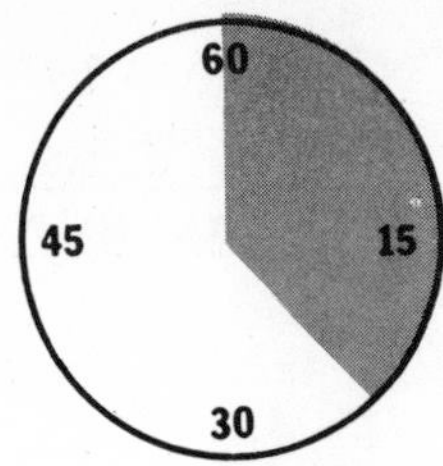

• 20 Seconds. Once a popular length. Now in lesser use. Usually the station charges as much for the twenty as it does for thirty-second commercial.

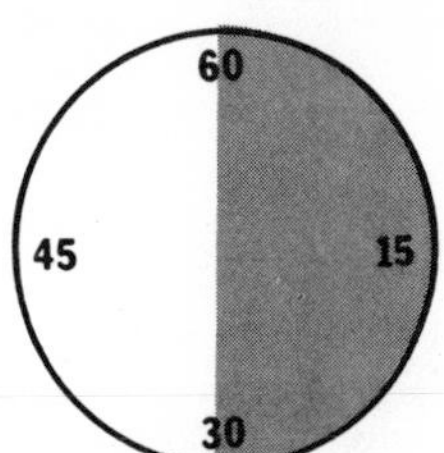

• 30 Seconds. This is the standard length of commercials. The price of the thirty becomes the base for other lengths. A ten-second commercial is generally priced at one-half the price of a thirty.

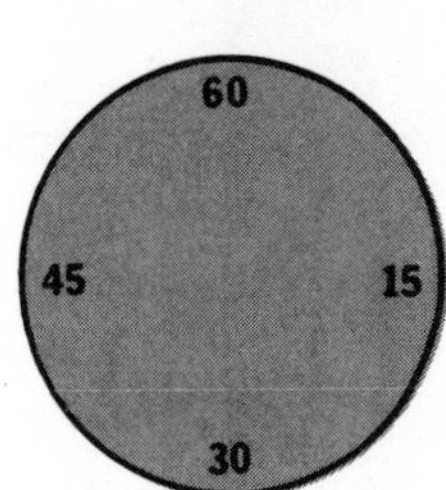

• 60 Seconds. For years the minute commercial was the standard length. With the coming of color, advertisers began to shrink to thirties. Where used, the 60-second commercial is priced at twice the cost of a thirty.

• 90 Seconds or Longer. Where an advertiser sponsors a show and "owns" the commercial content within that show, he might want to bulk his time and run fewer, but longer commercials. Some advertisers do this in a prestige show where they want to run the program without regular commercial interruption.

Local Rate Plans Vary from Station to Station

Even within a given market, stations may have different television plans and contracts. Usually the local rate is commissionable to the advertising agencies. It is advisable to confer with the desired stations to see which plans are available and which ones are best for your situation.

TELEVISION RATE PLANS

✓ Check Which Is Available

Annual Bulk Plan

This is like a newspaper lineage rate contract. You contract for 500, 750, 1,000 or more commercials annually. It allows you to schedule heavily at peak promotional periods and still save money at times when fewer spots are needed. ________

Example:

Rate Card WXXX-TV

Spots	500	750	1,000
Discount	5%	7.5%	10%

Frequency Discounts

If you agree to use spots during any continuous 13, 26, or 52 week period, many stations offer a discount for the frequency. ________

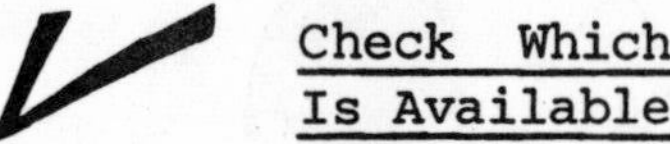

Check Which Is Available

Example:

Rate Card WXXX-TV Class C Announcements, 30 Seconds			
1 Time	13 Times	26 Times	52 Times
$100.00	$97.50	$92.50	$90.00

If you decide on daytime announcements, this rate card example shows it is "C" time and 30's cost $100 each. However, with frequency use of 52 weeks, the rate scales down to $90.00 each. This example is based on "C" time. Stations using frequency plans also have the plans for all segments of time, plus a mix of time segments.

Package Plans

When you buy a certain number of spots within a seven day period, starting on any day of the week, many stations offer package plans.

Example:

Rate Card WXXX-TV			
Class A 30 Sec.	5 Times $175.00	10 Times $165.00	15 Times $150.00
Class B 30 Sec.	$145.00	$140.00	$120.00

These package plans would apply to all of the time segments in addition to the "A" and "B" ones shown. Also, for different lengths of commercials besides the 30 second length.

Run of Station Plans (ROS)

This is similar to run-of-paper (ROP) with newspapers. Stations with this plan offer a package of commercials at a reduced rate with a station option to run them at any selected time. It is rare that an advertiser would know when the spots will run until the station "locks up" the station log for the next day. Some stations offer the ROS guaranteeing that a certain amount or percentage will run in different time parts such as so many in "A," so many in "B" and so on.

Buying ROS does not give control of locations of spots. This is satisfactory for commercials which are not dependent on the next day's selling, such as service and institutional messages. Also, they are good for "beefing up" a campaign for an event.

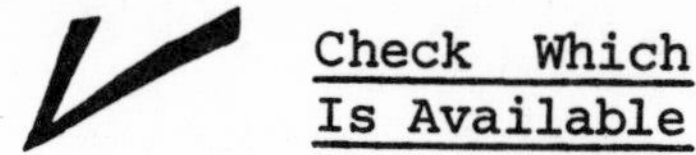

Check Which
Is Available

Example:

Rate Card WXXX-TV
Run-of-Station Rates

Twenty 10 seconds within a week $400.00
Forty 10 seconds within a week $700.00

ROS plans include other lengths besides 10 seconds.

Grid Plans

Because audiences vary in numbers within any time segment, some stations have changed their method of "A" time and "B" time for rates in favor of a grid plan. Throughout the entire broadcast day, they designate a code number to the specific time of day. In this way, if an early morning show had about the same audience as a mid-afternoon or early evening show, it would receive the same code number. To each of the code numbers, a price for the commercial is designated.

Example:

Rate Card WXXX-TV

30/20 SECONDS

CODE:	Section I	Section II	Section III
1	14	12	10
2	19	16	13
3	24	20	16
4	29	25	20
5	35	30	25
6	42	36	30
7	50	42	35
8	60	51	42
9	70	60	50
10	82	70	58
11	94	80	66
12	106	90	74
13	118	100	82
14	130	110	90
15	144	122	100
16	160	135	111
17	177	150	123
18	194	165	136
19	212	180	148
20	235	200	165
21	265	225	185
22	295	250	205
23	330	280	230

60 sec: double the 30/20 sec.
40 sec: 1-1/2 times the 30/20 sec.
10 sec: 1/2 the 30/20 sec.

Announcements purchased Section I are fixed and non-preemptible.
Announcements purchased Section 2 are preemptible on 2 weeks' notice.
Announcements purchased Section 3 are preemptible on 1 week notice.
Announcements purchased Section 4 are immediately preemptible.

In this rate card example, the code numbers, in a general way, apply to the usual A, B, C, D, etc., time slots which are generally referred to as prime, daytime, early fringe and late fringe. Also, there are three sets or sections of grid numbers which relate to whether the time cost is for fixed time or some variation of pre-emptibility. Preemptible and fixed prices are explained later.

Holiday Plans

In the same way that stores have special sales at different times of the year, some stations offer "holiday" plans which might be called by different names. These would occur at that time of the year when the station finds its inventory of spots larger than usual and offers what amounts to a clearance sale to sell them. Generally, the plans carry a definite discount from the regular plans. If a store already operates on one of the plans previously discussed, the holiday plan could mean an added discount.

Pre-Emptible Pricing

For each announcement, stations have fixed rates and pre-emptible rates. They are usually referred to as "sections." Some stations have rates in two sections, others in three or even four.

Section I: Fixed. This rate is higher than the pre-emptible ones. It guarantees a hold on the announcement and is sure to run on the designated time slot unless an unusual emergency comes about.

Section II: Pre-Emptible. This is a lower price for the same time slot. It is sold on the basis that if another advertiser wants to pay the full amount for the commercial, the fixed rate, the pre-emptible buyer loses out.

Sections III and IV. Some stations only use sections I and II. Others define pre-emptibility into additional parts. In these cases, they add the other sections to their rate card. A variation of three sections (I, II, III) is shown in the Grid Plan example. While the measure of pre-emptibility varies from station to station, usually Section II means that the lower-priced commercials can be "bumped" with two telecast notices. Section III, at a lower price per commercial, means that the campaign bought would be "bumped" with notice. Section IV, which is at the bottom rate, means that the commercials can be "bumped" without any notice at all.

In practice, most stations will inform a local advertiser as far in advance as possible about the danger of the lower-priced commercial being pre-empted. This is to give the good, regular customer the opportunity to shift its campaign or to change its buying for the campaign to a different pricing section.

In actual practice, most stations are fully aware of how their commercial inventory is at all times. Based on this knowledge, the station will advise an advertiser, well in advance of the advertiser's planning timetable, to use one of the sections at the higher price to guarantee that the commercials would run.

A New Pricing Structure

There are stations which approach pricing in a new way. On a continuous basis they examine their dayparts and individual programs throughout the full day, to see (a) what the audience content is and (b) what the demand is by the advertisers. Based on these two elements, they will adjust the price of the commercials.

For example, a specific show might be underbought by advertisers. The rate assigned to the commercials within it might be offered to everyone at Section IV. Any advertiser buying at that time is locked in at the Section IV rate. If the show picks up steam and selling of the time starts to climb, the station might increase the rate for the commercials within it to Section III or II or I based on the selling popularity. In this method of pricing, nobody is pre-empted. They own the spots in whatever Section they are bought. No one finds themselves "bumped."

Summary on Pricing Criteria

As we have seen, the price of a commercial is based on:

- Length of the commercial.
- Time of the day when it is aired.
- Size of audience.
- Fixed or pre-emptible rates.

Determining What You Get for Your Advertising Money

The measure of an advertising medium is its circulation. It represents the maximum number of people you can expect to reach. In newspapers, circulation tells the number of copies of the paper distributed. It does not tell anything about the number of people reading the paper or reading a particular advertisement.

In television, the number of television homes within the station's reception area is the television circulation, the number of households or people who can see your commercial.

In newspapers and television, not everyone sees every advertisement. In newspapers, there is a "readership" figure which tells what percent of the people reading the paper also read a particular advertisement. Studies have been conducted by Daniel Starch & Staff on newspaper readerships over the years. The measurements are based on:

- "Noted." This means the number or percentage of readers who saw or read any part of the advertisement.

- "Seen-Associated." These are the number or percentage of people who saw or read any part of the advertisement and were aware of the product or the advertiser.

- "Read Most." This is the number or percentage of readers who read 50% or more of the advertisement.

Newspaper readership scores, as determined by Starch, naturally vary according to the type of advertiser, such as department stores, tire stores, classified advertising and so on. Also, the by women-versus-men category is important for the different advertisers.

Newspaper advertisers could well afford to study Starch reports to get a better idea of what readership, out of the total newspaper circulation of a paper, it could expect to attract to their advertising. This is a true evaluation of the efficiency of a newspaper advertisement.

Newspaper readership scores vary according to the size of the advertisement. Naturally, full pages produce higher readerships than smaller sizes. However, the astonishing part of Starch readership studies is the proof that reducing the size from a page to seven columns, for example, hardly reduces readership at all. In fact, reductions of space, which means reduced dollars, are consistently less than the proportionate drop in readership size.

Some advertisers have cut back sizes of newspaper space per advertisement to use this money for a television budget and still hold, or nearly hold, all of their previous readership.

Television "Readership" and the Value of Ratings

In comparing newspapers and TV, the comparison should be between the number of people who read the advertisement, and not total newspaper circulation, and the number of people reached in television, meaning ratings.

Many things affect the number of homes reached by television. A look at a television station's coverage map shows where the signal can be received. Any station can also show the advertiser the latest television set count for its area. This is broken down by total households in the metro area, in the area where they have their greatest viewing influence and is their total coverage area. Fine break-outs in the audience demographics are also available.

The television audience is measured for each television market in the United States. It is paid for by the stations, by advertisers, and advertising agencies. Stations use this service to sell advertisers on the basis of the measured sizes of their audiences. Advertisers use the service to examine the demographic characteristics of their television audiences.

The size of audience tells the rating of the program or the time segment of the day. To this audience size, the price and value of a commercial can be compared in terms of the people reached and the audience.

There are two television audience measurement companies:

- The A. C. Nielsen Company
- The American Research Bureau

The television measurement reports which they publish are N.S.I. (Nielsen Station Index) and ARBITRON, sometimes called A.R.B.

Naturally, their measurement results per local market have some differences based on the method each company uses. This means that in examining audience measurement in your market, you must be sure to use the report of the same company to compare one station with another.

What Is a Rating and What Does it Mean in Time Buying

A rating is a percentage of the total television homes in the market viewing a specific station. A program with a 10 rating is viewed by 10 percent of all TV households. Ratings are used as a guide to show the percentage of the market reached.

- Total TV households multiplied by the rating gives you the number of households reached.

- Total women or men or teens in the market multiplied by the rating for each of these segments gives you the numbers of each of these groups reached.

- Total number of women or men in the subdivisions of the market, such as the metro area, multiplied by the rating of women or men gives you the numbers of each of these groups reached in that geographic area.

A High Rating Is Not the Only Answer

It is possible for a program to reach a large number of people, but they may be the wrong people. It might reach older people when your items are for young people, or children when you want men. You should consider the cost in relation to your rating.

The more interest the program holds for your target audience, as revealed in the ratings, the better your chances are for results. A large, total audience is great. However, in general, it is the target audience that counts and you can well afford to pay a higher cost per thousand to reach your target audience goal versus the total audience.

However, this must be weighed against the objectives and goals of the store and the merchandise per commercial. Target-audience goal is perfect to move the specific item, but you must consider the impression of the store that can be obtained by the non-target group in the total audience. For example, you might be selling jeans for teens and want to reach teens. However, parents pay the bills and they are also important in the TV coverage.

Other considerations outside of the specific ratings of the program are:

- Is a program personality selling for you?
- Does your message fit in with the atmosphere of the commercial?
- Does the program or the adjacent ones add authenticity to your message?
- Do the commercial facilities of the station mean a plus in the effectiveness of the commercials than do the ratings themselves?

A few rules to consider are:

- Do not buy time strictly on the basis of a high rating.
- Do not overlook excellent buys among the lower rated ones.
- Keep in mind the type of audience you wish to reach.
- See if you can find a program efficiently reaching this same type of audience.
- Evaluate the type of households you wish to reach in terms of family size, age of customer. Working women versus housewives might be another category.
- Select the time period that delivers the greatest proportion of this type of home for every dollar you spend.

The subject of ratings may make television time-buying sound complicated. Actually, it makes television advertising more scientific and efficient than other media. In newspapers you guess at the type and number of people who read the paper, read the section in which your advertisement falls and read your advertisement.

Ratings Versus Gross Ratings

The total of the ratings of all spots purchased in a campaign is referred to as Gross Rating Points, or GRP. Comparisons can be made between advertising weights of one campaign versus another. There are advantages of measuring advertising weights for campaigns and keeping them as guides. However, to be considered are:

- All campaigns do not require identical weight.
- Objectives and length of campaigns differ.
- Proper weights can be determined by trial and error or a constant evaluation of the weights that brought the desired results. Then this is applied to new campaigns.

The Rating (Audience Measurement) Services

The number of times a year the local markets are surveyed by both Nielsen and Arbitron is based on the size of the market.

ARBITRON™
Television

Audience Estimates in the
Arbitron Market of
Atlanta

SURVEY PERIOD: Sept. 19-Oct. 16, 1973
NUMBER OF TIMES PER YEAR THIS MARKET IS SURVEYED: 5

This report is furnished for the exclusive use of network, advertiser, advertising agency, and film company clients, plus these subscribing stations—
WSB WAGA WQXI WTCG

The "Total Survey Area" of this market is shown in white on the accompanying map. Where appropriate, the "Area of Dominant Influence" is indicated by coarse cross-hatching and the Arbitron "Metro (or Home County) Rating Area" by fine cross-hatching. Refer to the Glossary of Terms for complete description of these areas.

The example shown is the upper right corner of the opening page of an Arbitron Atlanta "book," as it is called in the television advertising industry.

The opening page of a "book" tells the period in which the audience survey was made as well as the number of times a year surveys are made. In the case of Atlanta, Arbitron measures and makes estimates of the market 5 times a year.

The report shows for whom the report was made. These are the people who paid for the survey. This Arbitron material is reproduced here through their courtesy.

An area map is shown on the same page. For the Atlanta market, and for each market "book," the map is an outline map which shows the central city and the counties surrounding it within and beyond the station's total coverage area. In the case of large cities in the area, they are also indicated.

Instead of an Atlanta map, or any market map, a prototype is shown here. It indicates, from the outside areas to the center, the total coverage of a station, the Area of Dominant Influence and the home county, plus adjacent counties.

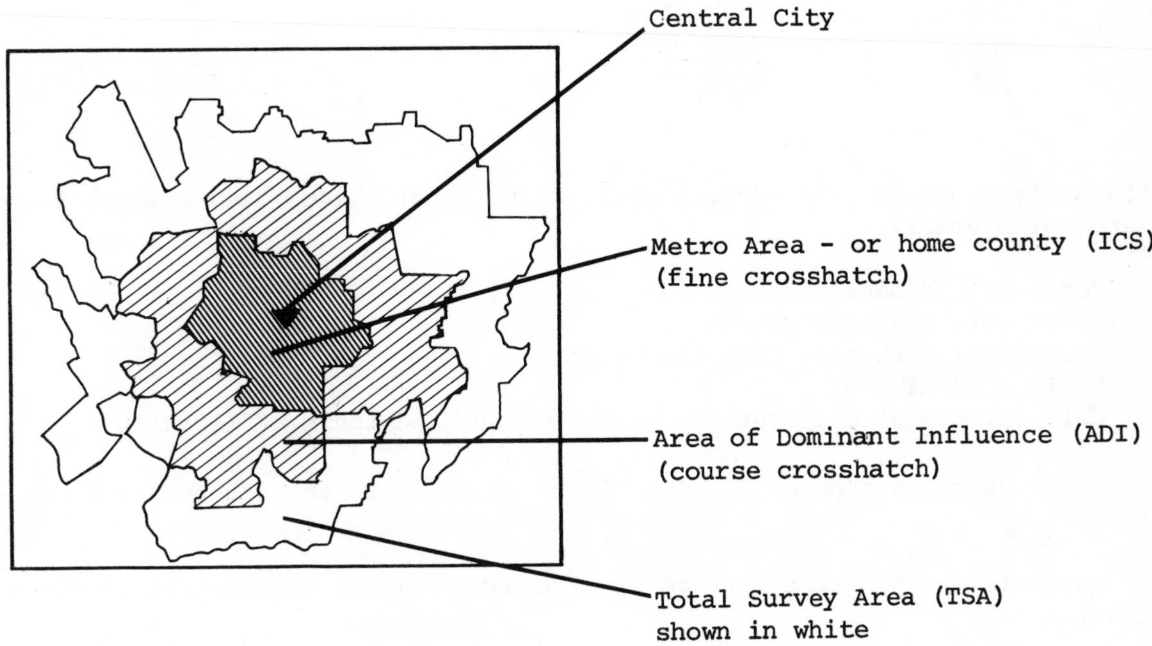

In the example shown:

- Shaded area is the metro area. Within the report, audiences are given for this area.

- Cross-hatch area is what Arbitron calls the Area of Dominant Influence. This is the area where viewers pay more attention to the survey stations than any other stations which may come into their homes from other cities. Separate audience figures are shown for the Area of Dominant Influence (ADI).

- White area. This is the total coverage area of the stations in the market examined. Separate figures are shown in "the book" for the total coverage area.

For the national advertiser with wide distribution of his products, the total coverage area is important.

The local retail advertiser has to evaluate his market and to learn where his customers are located. With most local advertisers today, the metro area is not his total customer market. The local advertiser is generally interested in the larger ADI area. Sometimes, depending upon multi-locations and good travel conditions, the local advertiser is interested in the total coverage area.

ESTIMATES OF HOUSEHOLDS IN MARKET

	TSA	PCT TV HH	ADI	PCT TV HH	METRO RATING AREA	PCT TV HH
TOTAL HOUSEHOLDS	1,568,400		770,800		548,000	
TV HOUSEHOLDS	1,519,600	100	751,100	100	537,200	100
COLOR TV HH	938,300	62	479,900	64	358,900	67
MULTI-SET TV HH	645,800	42	308,100	41	254,400	47
CATV SUBSCRIBERS	207,200	14	58,700	8	15,600	3
UHF TV HH (BASED ON INTAB SAMPLE)			656,600	87	494,200	91

TELEVISION STATION

CALL LETTERS	CHANNEL NUMBER	IDENTIFICATION AUTHORIZED BY FCC	AFFILIATION
WSB	2	ATLANTA, GA	NBC
WAGA	5	ATLANTA, GA	CBS
WGTV	8	ATHENS, GA	ETV
WQXI	11	ATLANTA, GA	ABC
WTCG	17	ATLANTA, GA	IND
WETV	30	ATLANTA, GA	ETV
WHAE	46	ATLANTA, GA	IND

In the Arbitron books, the opening page, among other things, shows total survey, ADI and metro areas:

- Total and TV households.
- Color sets and penetration.
- Households with more than one TV set.
- Cable subscribers.
- Call letters of all stations in the market and their affiliation.

The example above is from an Atlanta "book" as is shown through the courtesy of Arbitron.

Other pages in the Arbitron service give valuable market information in the ADI such as:

- Circulation of leading magazines.
- Circulation of the daily newspapers.
- Chain stores by name and by number of stores.
- Sales in total dollars by major classifications of stores.
- Population broken out by education levels.
- Occupations by white and blue collar workers, service people, farm workers.
- Population by families in different income levels.
- Value of housing.
- TV households in each of the counties.
- And a wide range of additional information helpful in any market decision.

<u>Information Found in Arbitron</u>

At each rating period, all stations in a market are examined from sign-on to sign-off for the rating period. Reporting is compiled in a variety of summaries. The typical "book" has a wide range of information. Major break-outs of audiences are shown throughout the pages with these as the most important.

- Program audiences.
- Daypart audiences with weekly summaries per station.
- Daypart audiences for each day of the week throughout the week per station.

In the Arbitron example shown, the names of the stations and the names of the programs have been blanked out in the example to keep them anonymous in this text. The example is the 4:00 to 4:30 afternoon period in an important market, averaged for four Mondays.

Share and ratings represent percentages. Audience numbers are in thousands. For example, an audience of "6" represents 6,000 viewers in that category.

WEEKLY PROGRAMMING | **4 WEEK TIME PERIOD AVERAGES**

DAY AND TIME / STATION / PROGRAM	ADI TV HH RATINGS WK 1 (39)	WK 2 (40)	WK 3 (41)	WK 4 (42)	ADI TV HH RTG (1)	ADI TV HH SHR (2)	ADI TV HH RATING TRENDS MAY '73 (43)	NOV '72 (44)	METRO TV HH RTG (3)	METRO TV HH SHR (4)	TOTAL SURVEY AREA, IN THOUSANDS: TV HH (5)	TOTAL PERSONS 2+ (6)	TOTAL ADULTS 18+ (7)	WOMEN TOTAL (8)	WOMEN 18-49 (9)	WOMEN 18-34 (10)	WOMEN 25-64 (11)	WOMEN 25-49 (12)	HOUSEWIVES TOTAL (14)	HOUSEWIVES -50 (15)	MEN TOTAL (16)	MEN 18-49 (17)	MEN 18-34 (18)	TEENS TOTAL (21)	CHILDREN TOTAL (23)	CHILDREN 6-11 (24)	ADI RATINGS: WOMEN TOT. (27)	WOMEN 18-49 (28)	WOMEN 18-34 (29)	MEN TOT. (32)	MEN 18-49 (33)	TNS TOT. (35)	CHD TOT. (37)
MONDAY																											4.30P						
	6	10		8																													
A			9		8	30	10	9	9	32	72	104	79	60	33	24	32	20	50	29	19	7	5	13	12	8	7	5	6	2	1	3	1
B	5	7	6	5	6	22	5	5	5	18	50	103	29	21	18	12	16	15	17	14	8	6	5	34	40	30	2	3	3	1	1	12	8
	4																																
		5	5	3	4	15	4	5	5	18	34	47	34	23	16	12	16	12	21	15	12	9	6	8	5	4	3	3	3	2	2	3	1
D	9		4	2	5	19	6	4	6	21	49	97	20	9	8	6	6	5	9	8	11	11	7	24	54	43	1	2	2	1	1	7	10
	2	3	3	2	2	7		1	3	11	19	40	4	2	2	2	1	1	1	1	2	2	2	4	32	20			1			1	7
	28	28	31	22	27		28	29	28		224	391	166	115	77	56	71	53	98	67	52	35	25	83	143	105	13	13	15	6	5	26	27
4.30P - 5.00P																																	

In the example, these are the points of importance.

- Day and time of broadcast -- the example shown is averages for a four-week period.

- Station call letters -- as indicated before, the call letters and the program have been purposely blocked out in this book.

- Ratings per week.

  Ratings in the TV households of the ADI (Column #1).

- Share of the households for each station (Column #2). This means the percentage of the total households viewing the particular station at this time of day.

- The ratings per station in the metro area (Column #3).

- The share of the households for each station in the metro area (Column #4).

<u>Audiences in the Total Survey Area</u>

- Total households viewing (Column #5).
- Total persons of two years + viewing (Column #6).
- Total adults viewing 18 and over (Column #7).
- Total women viewing (Column #8).
- Four different age break-outs of women (Columns #9, #10, #11, #12).
- Total housewives viewing (Column #14).
- Total men viewing (Column #16).
- Total men from 18 to 49 (Column #17).
- Total men viewing from 18 to 34 (Column #18).
- Teens viewing (Column #21).
- Children viewing in totals (Column #23).
- Children viewing in ages 6 to 11 (Column #24).

Audiences in the ADI

- ADI total women viewing (Column #27).
- ADI women 18-49 (Column #28).
- ADI women 18-34 (Column #29).
- ADI total men (Column #32).
- ADI men 18-49 (Column #33).
- ADI teens (Column #35).
- ADI children (Column #37).

Examining Specific Time Segments for Audience Composition

In the example shown, let us assume that stations A and B have the same charge for their commercials at the time shown of 4:00 PM to 4:30 PM. Let us assume that is the total coverage area in which you are most interested.

In Column #5 we see that:

- Station A reaches 72,000 households.
- Station B reaches 50,000 households.

From the standpoint of total households reached in the total survey area, obviously Station A is the best buy.

When it comes to the total number of people two years and older, Column #6 shows that both stations are reaching about the same number of people.

- Station A reaches 104,000 people.
- Station B reaches 103,000 people.

The next step is to see who are these people. In Column #7 we see that:

- Station A reaches 79,000 adults.
- Station B reaches 29,000 adults.

If it is total adults you want, obviously Station A is the better buy. But who are these adults? Column #8 shows us the women included in the adults and we see that:

- Station A reaches 60,000 women.
- Station B reaches 21,000 women.

This makes Station A a better buy for total women. In examining the breakouts of age groups in women in Columns 9-12 (which show a variety of age groups beginning at age 18), it is obvious that a good number of these women are in the 50+ age group. Even so, if it is the younger women you want, Station A is ahead.

When it comes to the teen audience, a different story is shown. Column #21 shows that:

- Station A reaches 13,000 teens.
- Station B reaches 34,000 teens.

The program column, as indicated before, has been blanked out in this text. Station B was definitely programmed for teens. This is also true of Station D which shows a larger teen audience than Station A, 24,000 teens in all. Station D also had a program of great interest to teens.

If the merchandise selected was targeted to teens, Station B was the better buy. With teen merchandise, adults pay the bills so consideration must be given to the adult audience in the time area.

Going back, if housewives were wanted mostly, Column #14 shows that:

- Station A reaches 50,000 housewives.
- Station B reaches 17,000 housewives.

Column #15 shows how many of these were housewives under 50 years. By subtracting Column #15 from #14, you see how many of the housewives are over 50. This composition must be considered based on the merchandise offered and the goals desired.

When it comes to men, Column #16 shows us that:

- Station A reaches 19,000 men.
- Station B reaches 8,000 men.

Columns #17 and #18 give the ages of these men. By subtracting, we see that a large number of men are in the 50+ age bracket, which is fine if they are the age group you want to reach. In any case, we see that on the last line in Column #16, which is the total of all the men viewers, only 52,000 are viewing television at all at this time of day versus 115,000 women viewing television at the same time (shown on the last line in Column #8).

This is understandable because working men are not available to view television in the middle of the afternoon. Nor are working women.

If, instead of the total survey area, you are only interested in the ADI area, Column #1 at the left and the columns on the right hand side are to be considered. The columns at the right give sex and age demographics. However, in other parts of the Arbitron book, even a finer demographic break-out is shown for age and sex.

In Column #1, we see that in the ADI:

- Station A has a rating of 8.
- Station B has a rating of 6.

When it comes to reaching men at this time, Column #32 shows:

- Station A reaches 2,000 men.
- Station B reaches 1,000 men.

This is the same pattern for men that we saw in the total survey area. In fact, if we add up all the men viewing television stations at this time of day in this market, Column #32, last line, we see that only 6,000 men were viewing television at all.

However, if the target audience is teens, Column #35 shows:

- Station A reaches 3,000 teens.
- Station B reaches 12,000 teens.

Station B also leads over Station A in children. We also see that Station D, with a program beamed at teens and children has more children than Station A or Station B and even more than both put together in this time period.

Up to now we have examined the audiences in the total survey area and the ADI area. If you are only interested in the immediate metro area, however, the same section of Arbitron gives you metro area statistics in Column #3. In general, most local advertisers are interested in the larger ADI areas.

In examining television commercial availabilities submitted by a station, it is possible to use the rating services and audience measurements to see exactly what audience you will get for your money. For total survey audience. For metro area. For ADI. For target audiences as well as total households. For different age brackets. By male versus female.

Cost Per Thousand As a Measure of Efficiency

In making comparisons between one commercial and another based on the cost, the term cost per thousand (CPM) is used.

> Cost per thousand is the amount in dollars that it costs to reach 1,000 of the measured audience. It is determined by dividing the audience in thousands into amounts of money.
>
> Examples:
>
> From the total survey area audience for the time 4:00 PM to 4:30 PM let us assume the price of each commercial is $100.
>
> Station A reaches 60,000 women, shown in the chart at 60.
> Divide 60 into $100.
> Result: $1.66 cost per thousand to reach women.
>
> Station B reaches 21,000 women, shown in the chart as 21.
> Divide 21 into $100.
> Result: $4.76 cost per thousand to reach women.

Station A reaches 104,000 total people, shown in chart as 104.
Divide 104 into $100.
Result: $0.96 cost per thousand to reach total people.

Station B reaches 103,000 total people, shown in chart as 103.
Divide 103 into $100.
Result: $0.97 cost per thousand to reach total people.

It is obvious that a request for avails should ask for information based on the exact total audience and target audience required. Stations are prepared to do this. Important words of caution:

- Make sure you compare the same audience composition for the same survey area when you compare stations to each other.

- Make sure the rating services are for the identical periods.

- Examine the program content of the rating period. It could be that one of the stations show an unusually high rating than it normally has in that rating period. A "special" show or an unusual feature could give the time segment a higher than normal rating for the station. The nature of the program will provide this tip-off.

To repeat, full information should be spelled out to the station(s) in asking for availabilities. A handy form is shown here. The station will report back on its suggested availabilities for the schedule wanted. A form typical of what stations use is also shown here.

*Arbitron charts and statistics courtesy of Arbitron (American Research Bureau).

REQUEST FOR TV AVAILABILITIES

FROM THE XXXX STORE

STATION ______________________ ATTENTION ______________________

PLEASE SUBMIT AVAILS BASED ON INFORMATION SHOWN BELOW:

NAME OF PROMOTION ______________________

SELLING DATES: FROM ______________ TO ______________

AIR DATES: FROM ______________ TO ______________

AVAILS SHOULD BE SUBMITTED BY: ______________________

CODE NO.	LENGTH	BUDGET	TARGET AUDIENCE	DAY PART PREFERENCE	AIR DATE	OTHER (Program Preference)

Include rating service information for each schedule giving total homes, total target audience, C/P/M for homes and target audience, and ADI rating.

TELEVISION STATION WXXX

ACCOUNT ______________________________

MERCHANDISE: ______________________ BUDGET ________

AIR DATE: STARTS _______________ ENDS _______________

NAME OF PROMOTION ______________________________

DATES OF PROMOTION ______________________________

NOTE: ______________________________

____

TARGET AUDIENCE

- ☐ WOMEN
- ☐ MEN'S
- ☐ TEENS M ___ F ___

- ☐ 18 to 34
- ☐ 18 to 49
- ☐ Under 18
- ☐ Under 12

SPOT INFORMATION					BUDGET/RATING/COST INFORMATION					
A Spots	B Length	C Day	D Time	E Program	F Cost	G ADI Rating	H Total Homes (000)	I Total C/P/M $	•J Target Aud. (000)	K Target C/P/M $
TOTAL SPOTS					TOTAL $	TOTAL Rating Points	TOTAL Impressions	TOTAL C/P/M	TARGET Impressions	TARGET C/P/M

COST (Column 'F') ÷ Target Audience (Column 'J') = C/P/M (Column 'K')

COST (Column 'F') ÷ Total Homes (Column 'H') = C/P/M (Column 'I')

Availability Form

The availability request form from the store for all five commercials is shown telling the station:

- Five different commercials. Although the specific merchandise is not indicated, in many cases the advertiser would tell the station what the merchandise is so that they would have a better "feel" and be able to do a better job in selecting the spots.

- Each commercial is 30 seconds.

- The budget per commercial.

- The target audience desired.

- Air dates to start and end.

- Additional preferences, such as the dayparts.

In this example case, the store is using two TV stations. The identical information went to both stations, if, as in this case, the store budgeted the same dollars on each station. If the dollars, however, were to be different, or the commercials different, the information to each of the stations would have been different.

The key in requesting "avails" is to give the station as much specific information as possible so the station can select and then offer the best spots to reach the target audience.

Note the bottom line on the form. In addition to target audience, the store requested information on total audience, knowing that the TV message would influence selling in:

- Areas beyond the ADI.
- Different age brackets for both men and women.

REQUEST FOR TV AVAILABILITIES

FROM THE XXXX STORE

STATION WXXX ATTENTION Mr. Jones

PLEASE SUBMIT AVAILS BASED ON INFORMATION SHOWN BELOW:

NAME OF PROMOTION Anniversary Sale

SELLING DATES: FROM 11/2 TO 11/7 inclusive

AIR DATES: FROM 11/1 TO 11/5 inclusive

AVAILS SHOULD BE SUBMITTED BY: 9/25

CODE NO.	LENGTH	BUDGET	TARGET AUDIENCE	DAY PART PREFERENCE	AIR DATE	OTHER (Program Preference)
Anniv-1	30	$500	Men 18-49	Part of Schedule Must be Directed to Working Women	11/1 thru 11/5 with 40% emphasis on opening day	No built-in program preferences. Goal is to reach the right audience for the campaign.
Anniv-2	30	$450	Women 18-49 Men 18-49			
Anniv-3	30	$450	Women 18-49			
Anniv-4	30	$500	Women 18-49			
Anniv-5	30	$550	Women 18-34			

Include rating service information for each schedule giving total homes, total target audience, C/P/M for homes and target audience, and ADI rating.

Time Buying Example

With the form, let us take an example of buying time.

Market

Consists of a group of closely located cities and towns.

- Metro area: 75,000 households.
- ADI area: 230,000 households.
- Total survey area of TV stations: 490,000 households.

Retailer

The retailer is a specialty department store with several locations. They are located, not just in the metro area, but throughout the TV stations' ADI area. This is the basic location of their customers. The balance of the TV coverage, the difference between Total Survey Area (TSA) and Area of Dominant Influence (ADI) is not "waste" circulation because many families in that area can travel to one of these stores on good roads.

The Event

An Anniversary Sale lasting a week: 11/2 through 11/7.

Merchandise/Commercials

Five commercials were prepared. Each commercial had the same "intro" which sold the Anniversary event. In each of the five cases, specific merchandise followed the "intro."

- One commercial was on men's suits. Primary target -- men 18 to 49.

- One commercial was on men's shirts and shoes. Dual target -- men 18 to 49. Also women 18-49 because women do a lot of the buying of men's shirts as well as influencing the man's decision on men's shirts and shoes.

- One commercial was on domestics: sheets, cases, bed spreads. Primary target -- women 18-49.

- One commercial was on young boys' and girls' clothes. Primary target -- women 18-49 because they do the basic buying on this merchandise.

- One commercial was on misses pant suits. Primary target -- women 18-34. This target was selected even though the store knew that older women also buy these specific styles of pant suits. However, this buyer considered the 18-34 women audience as the major target and buyer of the pant suits.

Budget

The budget for the entire campaign was $5,000. This averaged $1,000 per commercial. However, some of these commercials, because of the merchandise, deserved somewhat more money and others, somewhat less money. Actually, the $5,000 is a gross budget. Four out of the five commercials had co-op dollars. This amounted to about $2,000. The net budget of the campaign for the store was only $3,000.

From past experience, the store knew that a budget of $500 a commercial per station would give them a schedule that would deliver about 100 total gross rating points, provided the spots were properly scheduled. With the commercial on both stations in this market, giving each station approximately $500 each, the total, or G.R.P., would about double, meaning about 200 gross rating points. In this case there were five items and five commercials per station. If properly selected, the five commercials were expected to reach 500 gross rating points per station for the entire campaign, or 1,000 G.R.P. for the two stations in this campaign. This is heavy saturation.

Schedule Strategy

With this budget of $5,000, the strategy was to come on with about 40% of the budget at the very start to create momentum and then to level off. Even eliminate TV the day before the close of the sale, leaving the "Tomorrow -- Last Day" theme for the newspapers.

The five commercials, while running on the same days, would require spacing throughout the five days so they were not "back to back." For this purpose, as a check, a composite schedule would be put together to see all of the time spots to be sure there was not too much concentration at one point nor too spotty of a scheduling format. A form for this master schedule is shown.

Two television stations are in the market. The plan was to give each station about half of the budget.

Request for Availabilities

The "avails" were requested of each station for each of the five commercials. Both stations were told they were being considered. They were given target information on a store form as shown. This included:

- Air dates.
- Type of merchandise per commercial.
- Target audience per commercial.
- Budget per commercial.
- Preferences for dayparts, etc.
- All five commercials were to run on the same days with proper spacing of time from commercial to commercial so that there would be no back-to-back or "thin" scheduling times.

Recommended Station Schedule

From the availability request from the store, the station would select spots which match the requirements. In this case, the filled in form is shown. Primarily, it presents for each of the five commercials:

- Air dates for start and end.
- The target audience: age, sex.
- The day and time of each spot.
- The program in which each one is located. (The store might have some objections to having specific merchandise within certain programs.)
- Cost per commercial.
- Ratings per commercial.
- Number and cost per thousand of total households reached.
- Number and cost per thousand of ADI households reached.
- Total number of spots with total cost for this commercial.
- Total number of rating points. This is generally referred to as G.R.P., gross rating points. Note how, for this item and station, the schedule achieved 115 gross rating points.
- The size of the total audience (in thousands). In this case, 266,000.
- The cost per thousand to reach this total audience. In this case, $2.08.
- The size of the target audience reached. The box in the upper right corner shows the target. In this case, women 18 to 34. The total schedule on this item reached 84,000 of these women.
- The cost per thousand to reach the target audience. In this case, $6.61.

The form shown here is for the pant suits, one of the five commercials which made up the store's anniversary sale. A <u>similar form</u> would be submitted by the station for <u>each</u> of the five commercials.

TELEVISION STATION WXXX

ACCOUNT The XXXX Store

MERCHANDISE: Pant Suits for Misses BUDGET $550.00

AIR DATE: STARTS 11/1 ENDS 11/5

NAME OF PROMOTION Anniversary Sale

DATES OF PROMOTION 11/2 through 11/7

NOTE: One of five commercials in total campaign for the Anniversary Sale. Same air time.

TARGET AUDIENCE

☐ WOMEN

☐ MEN'S

☐ TEENS M ___ F ___

☐ 18 to 34

☐ 18 to 49

☐ Under 18

☐ Under 12

SPOT INFORMATION					BUDGET/RATING/COST INFORMATION					
A Spots	B Length	C Day	D Time	E Program	F Cost	G ADI Rating	H Total Homes (000)	I Total C/P/M $	J Target Aud. (000)	K Target C/P/M $
1	30	11/1	2:00 P	As World Turns/ Guiding Light	$43	15	34	1.26	10	4.30
1	30	11/1	3:30 P	Price Is Right	$26	11	25	1.04	9	2.89
1	30	11/1	9:00 P	Mash/Mary Tyler Moore	$132	21	49	2.69	16	8.25
1	30	11/2	3:30 P	Price Is Right/Match Game	$26	11	25	1.04	9	2.89
1	30	11/3	9:00 P	Mash/Mary Tyler Moore	$132	21	49	2.69	16	8.25
1	30	11/5	2:30 P	Guiding Light/ Edge of Night	$34	11	26	1.31	7	4.86
1	30	11/5	9:00 P	Gunsmoke/Lucy	$162	25	58	2.79	17	9.53
TOTAL SPOTS 7	30's				TOTAL $ $555	TOTAL Rating Points 115	TOTAL Impressions 266	TOTAL C/P/M $2.08	TARGET Impressions 84	TARGET C/P/M 6.61

COST (Column 'F') ÷ Target Audience (Column 'J') = C/P/M (Column 'K')

COST (Column 'F') ÷ Total Homes (Column 'H') = C/P/M (Column 'I')

Analysis of the Proposed Schedule

In the schedule submitted, the C/P/M for target audience for women 18 to 34 per spot ranged from a low of $2.89 to a high of $9.53. For total homes, $1.04 to $2.79 C/P/M per spot or a total schedule of $2.08 C/P/M for total homes reached.

Within this range, these commercials in daytime ranged from $2.89 to $4.86 C/P/M for the target audience for women 18 to 34. The two commercials at $2.89 are obviously much better buys. The ones at $4.30 and $4.86 are not. Perhaps these two should have been shifted to other times on the station log to bring better efficiency. On the other hand, perhaps the two at $2.89 C/P/M each represented exceptional buying opportunities. This is a point for discussion by the store with the station.

The higher C/P/M commercials of $8.25 and $9.53 are in prime time. Naturally, they are a higher cost per each because total audience is much greater. Column "H" bears this out. The last commercial on the list (11/5 at 9:00 PM) delivers a total of 58,000 households, of which 17,000 are the target 18 to 34 women.

To decide if the big difference in dollars between these prime time commercials is worth it, the store, in this case, knew:

> Almost 50% of women in their area are working women. The money they earn adds to the total income of their family. This makes these women better customers. Also, many working women think of their earnings as their money, not primarily family money. That makes them better customers. It is obvious that it is worth more to spend more money to reach better customers. In this case, the working women can only be reached during prime time.
>
> The target was for women 18-34. At the same time, the store knew that many women 34-49 were good customers for this specific item. The store could have asked the station to refigure the C/P/M for women 18-49. For this schedule, using the same commercials at the same time the C/P/M for 18-49 women would have amounted to only $2.49 instead of $6.61. To repeat, total houses reached brought the C/P/M down to $2.08.
>
> The target was for women 18-34 knowing that women 34-49 included some good prospects. But what about the men who saw the prime time commercials? The percentage of these men who influenced women to buy the specific merchandise, misses pant suits, was probably small. However, the total commercial had a selling effect on men. It told them about the Anniversary Sale at this store. It must have impresed many of them to (a) go to the store for other things during the sale, or (b) suggest to their wives to shop there, or (c) give them an impression about the store's values and influenced them for their future shopping.

Some Conclusions

Ratings and target audience for each commercial are important to achieve efficiency in time buying. However, as pointed out before, it is necessary to look beyond the ratings to see if other things are important too. In this specific case, we see that women outside of the target age benefit the store and that the message to the men brought the store benefits, too.

Competitive Station's Schedule Proposals

Examining the schedules, such as shown here, with the schedules proposed by competitive stations, gives the advertiser the opportunity to compare the cost efficiency of one station with another.

Using One Station or Several

No retailer likes to scatter his buying of any classification of merchandise over a wide range of resources. This would make the retailer of little importance to any one of the vendors. Instead, the typical retailer holds merchandise buying per classification down to a limited number of resources.

For similar reasons, if the budget is a small one on television, it is probably best to confine the schedule to one station. As the budget grows, other stations can be added.

Making a Master Schedule

Stations submit their avails and the final confirmation of the orders in different ways. A separate form should be used for each item or commercial so that each one stands on its own in any given campaign.

However, if there is more than one commercial and more than one set of items in the same campaign period, it is best to create a master schedule as a check to be sure that the commercials do not all bulk up at one time or day, unless that specific effort is desired.

For this reason, a master schedule becomes a convenient way to examine the total schedule in a campaign and the total impact upon the viewers of the station or stations used. Commercials should be listed in chronological order for the day and time of the day. This total schedule is one which can be distributed to all executives and all sales people involved in the merchandise offered on television. A form is shown for guidance:

MONTH OF ____________________ STATION ____________________

WEEK	PROMOTION OR ITEMS	DAY OF WEEK	PROGRAM	TIME OF DAY	DOLLARS

TIME BUYING CHECK LIST FOR EVERY TV CAMPAIGN AND EVERY ITEM IN THE CAMPAIGN

Check Each Point Accomplished and Indicate Date of Decision

1. Determine campaign objectives. ________

   Who you want to reach? ________

   When? ________

   Budget? ________

2. Tell your TV stations of your objectives. ________

   Immediate cash register results or long range? ________

   Tell them your budget -- as far in advance as possible but at least three to four weeks ahead. ________

   Ask for availabilities. ________

   Specific audiences information on the people you want to reach by age and sex. ________

   Ask for the cost efficiencies for each announcement submitted. The number of homes and cost per thousand or number of target audience and cost per thousand. ________

3. Select from the announcements submitted. ________

4. Examine the audience information and costs to be sure your objectives and budget are met. ________

5. Have the stations give you a confirmation. Check for this: ________

   (a) Day of each announcement. ________

   (b) Time of each announcement. ________

   (c) Cost of each announcement. ________

6. Give stations written instructions on what commercial is to run in which time indicated. ________

7. The stations will give you an affidavit of performance and a bill. Check the bills to see that all announcements ran at the agreed time. There are times when a shift of time is needed in which case the station offers you a "make good." This is another announcement equal in audience to the one ordered. ________

DECISION #12

Get Full Advantages of Television: Tie-ins With Other Sales Promotion Media

Most advertisers believe in coordinating all their sales promotion media. In this way you squeeze all the juice out of the sales promotion orange.

Stores believe in the power of repetition. For special events, stores use the same signs and posters throughout the store and in the windows. When the television power of repetition can be harnessed to reach its vast audience, the effect is synergistic.

Coordinating Television With Other Media

Sales promotion is the coordination of all selling activities to obtain profitable sales. If an item is good enough to be advertised, it should be promoted all down the line. This coordination increases the impact through constant repetition and presentation of the item, the event, the idea or service.

The glamour of television offers promotional tie-in possibilities beyond traditionally used, retail advertising tools.

Who Is Assigned To Follow Through?

Store Window Displays

A window with "today's television special" is one way stores use a tie-in. People passing the store recognize "here is the dress I saw on television" and react to it. Displays can include, of course, other items that are related to the television special. For example, accessory items including hats, handbags, gloves, and shoes can be shown with dresses. Or if the television feature is a power lawn mower, the store window can also show other gardening items such as rakes, seed, fertilizers.

____

In addition to related items, the television special window can feature higher-priced lines to trade up the customer. Or, if the item is one which people might buy in multiples, such as shirts, blankets, hosiery and others, the item can be displayed with a multiple price, with the opportunity to charge it to a credit account.

____

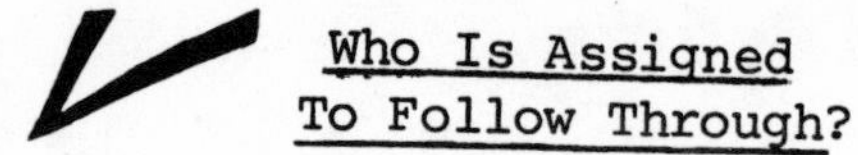

Who Is Assigned To Follow Through?

The store windows can also show posters of the television personalities connected with the store television campaign and other television stars at the station itself.

Interior Store Displays

Many of the ideas for store windows can be used in interior store display.

In the section of the store where the item is regularly sold, many stores create a feature display built around the television specials. The handling of the display depends on the nature of the items. If they are demonstrable, a top salesperson can show the customers the way the item works, or is worked, or is used. For example, if the item is a cosmetic, an attractive girl demonstrates the product the way often seen in drug and department stores.

In addition to feature displays of television specials in the usual departments, there can be displays in other departments as well. In this way, sections of the store take on a new look and add extra interest. Depending on store policy, the items can be sold in both sections or posters can direct the customers to the regular selling department.

Sign Toppers

For added identification and reminder, a good idea is to create sign "toppers" for the television items. In this way, customers identify them quickly. Their use in a store causes customers to look for the television attractions when they shop.

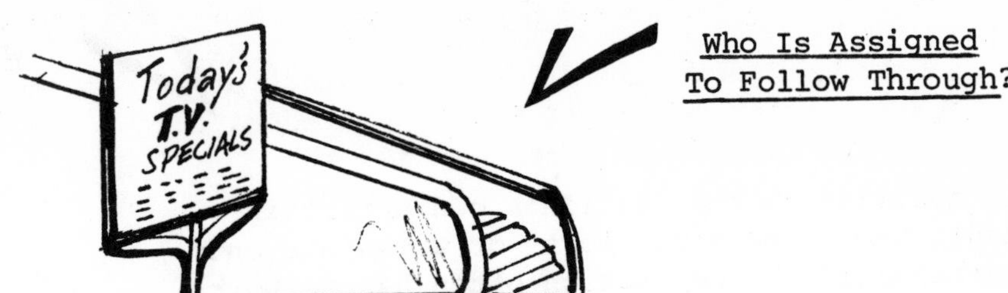

Who Is Assigned To Follow Through?

Sign at Elevator and Escalator Locations

Signs adjacent to the elevator where people wait and at the foot of escalators are locations used by stores to promote their television specials. The poster could be a permanent one captioned "Be sure to see today's TV Special." The poster could be slotted so that strips could be inserted daily with the current special. __________

Personal Appearances of TV Personalities

People like to see and meet television personalities and this includes the people who do commercials, too. They produce tremendous store traffic. A children's show personality, the local cowboy hero, or TV circus clown is a drawing card for children with their parents who will stop at a children's department, toy department, pet shop, sport shop or any store with children's interests. __________

A television personality, readying for an appearance in the store's auditorium event, whether it is a cooking school, gardening classes, home furnishings or interior decoration classes, is an attraction for increased customer attendance. __________

There are many cases where stores have been able to have their on-air television salesperson actually behind the store counter. When this is promoted, the idea produces extra traffic and extra sales and serves to enthuse the regular sales staff. __________

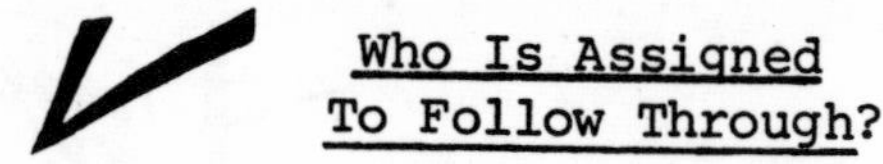

Who Is Assigned To Follow Through?

In addition to the station's own entertainment, news, weather, sports or women's events personalities, the station often has network talent visiting their stations. Establishing contact with the station can result in using these people for store traffic.

These stores have their own television star address women's clubs, appear at fashion shows, schools, hospital entertainments, children's and old folks' homes and numerous other public functions. This is a case of using the popularity and glamour of television to attract a larger audience of customers.

Store Salespeople's Enthusiasm

Some stores hold meetings in which the entire television campaign plan is explained to salespeople. There are general or departmental meetings. The purpose is to get everybody into the television act. Salespeople are enthused with things that are new, different, and exciting. Television is certainly all of these. At meetings like these, stores often bring the television personalities into the store meetings to add extra "show biz."

DECISION #13

Keeping Records for Future Promotions

Retailers keep a scrap book of their newspaper advertisings. The store does this, not just as an historical record, but as a guide for future advertising. In fact, most stores keep scrap books of their competitors. From their competitors' scrap books they know when their competitors plan all of their major events: when they begin the back-to-school campaign; when they break with Christmas promotions and the Santa arrival; when the anniversary sale begins and so on.

In the store's own scrap book many things are customarily added: When the advertisement ran. What the weather was. What the cost was. The sales of the items promoted as well as the departmental sales. If the advertisement is a repeat of a recent event or a repeat of an event of one year ago, these comparative figures are included. All this is done as a guide for new planning.

Television does not have an easy way of doing a scrap book of commercials. The naked eye cannot see a film print or a tape. Besides, the tape or film is too long to reproduce on paper. But there are simple, easy ways to prepare a TV scrap book in a continuous way to have the documentation on what happened and the results.

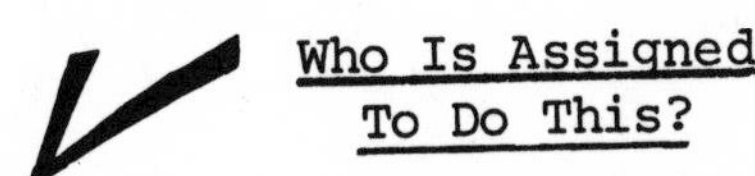

Who Is Assigned
To Do This?

A regular three-hole note book can be used. Each spread of pages would give the actual schedule by time periods for each commercial in a given period. Stations. Costs.

____

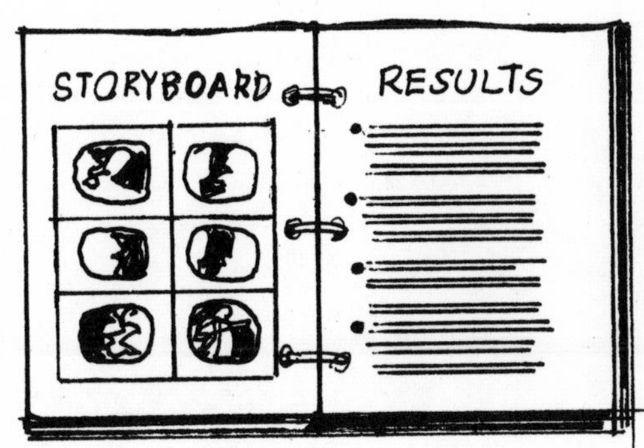

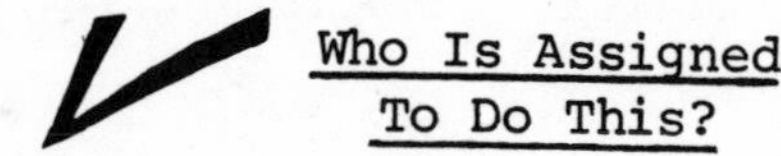

Another page would show the actual script. This would be followed by a condensed storyboard. If no storyboard was used in actual production, it is possible to shoot Polaroid photos of the commercial while in production or to enact the basic scenes subsequently in studio. If this is not practical, simple "stick" drawings may convey the idea.

The final page would be for sales results plus any comments that would be valuable when deciding if the commercial should be used in full or in part at a future time.

Today the videocassette is becoming more and more popular as a part of the store's equipment. It is frequently used in store training departments. A videocassette enables the store to transfer film or tape to a convenient book-sized package and store on a book shelf. On the side of the package is a place to record what the commercial is and when it ran. On the inside is plenty of room to add specific information about the commercial, the budget, schedules, sales results and other details.

SPONSOR A PROGRAM OR USE ANNOUNCEMENTS -- OR BOTH?

In the early days or television, similar to the early days of radio, manufacturers sponsored programs. A weekly show sponsored by a single manufacturer was typical. The same people would set aside the special time required to follow the show each week.

Television audiences grew. With the larger audiences, the rates increased. Manufacturers became a part sponsor of a show sharing the show with a non-competitive advertiser. Even on such shows, a local station often cuts other advertisers in on the station breaks. Sharing a show gives the advertiser the added advantage of spreading his money. A time budget for one show can be cut in two parts and used for two shows. This gives the advertiser the opportunity of reaching more people in total as well as reaching some of the people twice. Of course, many manufacturers like the idea of running seasonal "specials."

Advantages of a Sponsored Program for the Local Advertiser

Certain advantages exist for the sponsorship of a regularly scheduled show, or the co-sponsorship of one:

- Gives prestige to the advertiser. He is important. Has his own show.
- Builds a loyal audience, a following. The viewers tend to see several of your commercials within the show.
- Program personalities can often be used in your commercials.
- Program personalities can often be "merchandised" for personal appearances in the store or at other store activities. The TV personality becomes the store spokesman.

Disadvantages of a Sponsored Program for the Store

- The program sponsorship often includes the cost of talent as well as the commercial time.
- With the same dollars it is possible to reach more people, more often by using TV spots at determined times.

- Available local shows might be at the time of day not suitable for the store's target audience.

- The show tends to build up a regular following. The same people watch each time. With the same dollars you reach more and different people by selecting appropriate spots for your target audience.

Advantages of Spot Announcements Bought Separately

- Flexibility in the number of commercials used by day, week or month.

- Flexibility in placing commercials at the time of day to reach specific audiences.

- TV spots can be cancelled with a reasonable notice. However, shows require commitments of 13 or more weeks.

Advantages of a Sponsored Program Plus Spots

Where a budget is substantial, a combination of the two methods of using TV time becomes advantageous because the plus features of both are put into action.

Of course, there are various local programs that are in opportune time slots which reach maximum audiences. News shows are a typical example, which is the reason many retailers, banks, utilities and other local advertisers like this kind of local program.

WANT TO KNOW MORE ABOUT TELEVISION? OTHER PEOPLE CAN HELP YOU!

Any television "pro" can tell you that there is much help available for the experienced as well as the beginner in television advertising. Here are some of the places where you can turn and all of them are free:

Advertising Agencies

Most retailers handle their own newspaper advertising because they feel they are better equipped to prepare their day-to-day advertising then an outside agency.

Many stores use advertising agencies or specialized advertising companies in specific areas. Direct mail catalogues and a special campaign, such as an important anniversary sale, are examples.

Television is also a specialty. It is an area where many stores have limited knowledge. For this reason, it is often valuable to enlist the assistance of advertising agency pros. In cities all over the country, there are agencies interested in retail business. These are the agencies to contact.

An agency can be employed to do the entire television advertising job or can be used for just certain segments of it. This depends upon the store and the advertising campaigns involved. An agency gets its fee from the TV station as a commission on time charges.

Production Houses

These are studios which have the expensive equipment for television production. Equipment does not vary radically from studio to studio. However, people vary. It is wise to select a production house with good directors, a house which is willing to give you the time and the energies needed. This is especially true for those stores new to television.

Production people like to share their knowledge. Visit their facilities. Learn by seeing what a studio looks like and does.

Local Directors, Writers, Artists

These people are not affiliated with any specific company and take on jobs for advertisers. Generally, they know the good local and nearby production facilities and can be extremely helpful with short cuts and time saving.

Rating Services

These are paid services, published several times a year, which give the audience of your local television stations by quarter-hour breaks through the day. All demographics are given of the audience. This includes breakouts by age and sex so you can select the spots which match your target audience.

Model Agencies

For television on-air talent there are model agencies. Some stores employ actors from "little theater" groups in their city. In general, it is more efficient to employ professionals rather than to rely upon store people.

Vendors

Some vendors have complete films showing their products in action. Many of these films have openings for you to add your own store's name and location. Some vendors supply scripts, prepared by professional television writers, which can be used or modified in making your own commercials.

Some vendors provide original art work. It can become part of your store's commercials.

Many vendors offer cooperative advertising allowances. These vendors know television's ability to sell their merchandise and they want you to use television for the same reason. Many stores instruct their buyers to ask for co-op dollars and vendor television aids when they are interested in buying trips.

The Stations Themselves

Television stations recognize that television is a medium new to many stores. They assist stores in:

1. Market area studies

   Stations can examine the market area of the retail advertiser and show how it can meet those market requirements.

2. Time of day for telecasting

   Stations will study the target audience wanted by the store and recommend the best time to reach these customers. This depends on the type of customer, age, sex, and income level.

3. Budget for television

   Based on the store's volume, sales potentials, present promotion budget, competition and other details, stations can offer recommendations for a store's television budget and length of a campaign.

4. Music libraries

   Stations have music libraries for the store to use. There is usually a small fee.

5. Artwork

   Many stations have facilities for creating artwork for the store. Supers. Backgrounds. Slides.

6. Copy

   Stations are prepared to advise on copywriting. Many have copy staffs. They can write actual copy based on information provided by the store.

7. Program production

   Prop rooms are maintained by almost every television station. They include a wide variety of usable props which are usually available to stores. However, it is best for stores to consider their display departments for prop sources.

Buying Offices, Trade Associations

The store's own buying office should not be overlooked. These offices have case histories on what "sister stores" are doing on television. In some cases, it is possible to secure commercials from other stores and adapt them.

Trade associations such as the National Retail Merchants Association have files on what other stores do. Commercial examples, too.

The Television Bureau of Advertising with offices in New York, Chicago, Detroit and Los Angeles has examples of thousands of commercials used by other stores. These can be examined for ideas. Case histories. Commercials. How-to-do-it booklets. One Rockefeller Plaza, New York, N.Y. 10020.

But the Biggest Way to Get Help Comes from Two Do-It-Yourself Directions

- Get into a studio and see the action. See how they actually create a commercial. See what goes on and can be done in the control room. Get behind a camera. Learn by watching and asking questions.

- Study commercials. When you watch television, forget that you are casual viewers. Examine each commercial. See the basic idea in the commercial. Let your imagination go and determine how the identical idea can be switched into an idea for your own company.

LIMITATIONS OF TELEVISION

Newspaper advertisements can be read at any time and at any place. Different members of the family can read the advertisements at different times or they can read different sections of the paper at the same time. Television advertising can be seen only by customers when the commercials are on the air.

With television it is important to buy the time "slots" which reach the audience you want to reach. To insure catching them, repetition of the same commercial at different times of the day is essential. In addition, this gives you repetition for much of the audience. Repetition in television is called frequency.

Newspaper advertisements can be big or little depending upon the message you want to deliver. A small advertisement might be enough for a relatively small message, such as one item. A page or double truck can be used when the even is of big importance. A section in the paper can be used for more important events.

With television, this impact can be achieved by running the same advertisement over and over again at different times of the day or on different days. Or, different commercials can be used in the same period. For an event, it could be many commercials that have the same opening, an opening which describes and sells the event. This opening can be followed by different merchandise in each or in several of the commercials.

Newspaper advertisements can contain many items. In fact, the newspaper advertisement can be a catalogue.

With television this is not possible. Three items is about maximum for a commercial and they should be related in merchandise or price or some other "hook." This means that in television it is needed to edit down to the most important items to advertise. This is also good merchandising.

Newspaper advertisements can be torn out by a customer and taken to the store as a shopping guide.

With television this is not possible. Some day in the future there may be a print-out method. The TV commercial, using the eye and ear with music and dramatic effects, must have the proper impact to be memorable.

Newspaper advertisements can be a tear sheet. These can be used for scrap books serving as a record and reminder. The scrap books are kept as a guide for future planning.

> With television, the same can be achieved with a scrap book of scripts or storyboards. Even Polaroid photos can be taken of the commercials in studio. Along with schedules and costs, these can be used as a TV scrap book to record results.

Newspaper advertising tear sheets can be used for co-op records. They serve as proof to the vendors that the advertisement ran.

> With television, stations furnish an affidavit of performance. This, with a copy of the script, becomes the same thing for the vendor.

Television vs. Sales Promotion

Newspapers are not the total answer to the sales promotion plans of any advertiser. Nor is television. Nor is radio. Nor is direct mail. The advertiser must decide what kind of advertising and what media is needed in a sufficient advertising mix to reach all of its potential customers with the advertising budget and its command.

UNDERSTANDING THE TELEVISION LANGUAGE

Every industry has a language of its own. Television language has borrowed from the theater, motion pictures, radio, and advertising language. To these, it has added many new words and terms. The TV advertiser needs to know the language in order to communicate with stations and production people. Shown here are the most popular terms.

- A -

A and B ROLL. With film, "A" shots on one roll and "B" shots on the other. Negative, black leader is alternated. They are then combined as one film. With tape, two tapes are fed into electronic editing to combine as one tape.

ACROSS-THE-BOARD. Television activity scheduled five or six days a week at the same time.

ACADEMY LEADER. The start of a film showing a countdown backwards from ten seconds to where the film actually begins. Allows projectionist to properly focus.

ABSTRACT SET. Neutral background for the set.

ADJACENCIES. Programs or announcements immediately preceding and following a specific time period on a station.

AD LIB. Impromptu action or speech not prepared or written into the script.

ADI - AREA OF DOMINANT INFLUENCE. An area of counties where the home market TV stations are dominant in total hours viewed. Stations from other markets might be viewed, but are not dominant. A term used by Arbitron. It is similar to DMA used by Nielsen.

AFTRA - AMERICAN FEDERATION OF TELEVISION AND RADIO ARTISTS. The union primarily for talent performing on tape.

AFFILIATE. A station affiliated with one or more networks and which can carry shows and commercials originated by the network.

AIR CHECK. A video tape of a show as it is aired.

ALLOCATION. The power and frequency given to a station by the F.T.C.

AMBIENT LIGHT. General studio light, not specifically on the main subject.

ANIMATION. A cartoon-type technique to give motion to still artwork.

ANNOUNCEMENT. A commercial, station "promo" or public service announcement.

ANSWER PRINT. The original print from a negative film. It is examined for possible color and quality corrections. When accepted, it is called a release print.

ARC LIGHT. Electricity passing between two electrodes creates the arc light.

ARB - AMERICAN RESEARCH BUREAU. Known as Arbitron. Examines markets, determines demographics and rating points.

ASPECT RATIO. Ratio of picture, width to height, 4 to 3 in TV and films.

AUDIENCE COMPOSITION. Breakdown of men, women, teens and children in TV audience at a given time.

AUDIO. The sound part of TV.

AVAILABILITY. Open time suitable for sponsorship on TV.

- B -

BAR - BROADCAST ADVERTISERS REPORTS. A company which measures television advertising activity in 75 markets. Reports length and frequency of commercials and estimates of dollars spent.

BACKDROP. Curtain or flat which can be moved out of scene.

BACK LIGHTING. Lighting of performer from behind. Gives depth to scene.

BACK TIME. Timing from the end instead of the start of a show.

BALOP. Short for the baloption machine which projects objects, photographs or stills.

BARN DOORS. Adjustable sides or top shades that fit over lights to narrow or widen light beam.

BG. Background.

BILLBOARD. Credits at opening and/or closing of program listing advertiser, talent, producer, director, writer, etc.

BLAST. Extreme sound distortion.

BLOOP. Removing sound from a track.

BLOOPER. A mistake.

BLOOM. White spot with black areas at edges caused by objects, usually shiny metal, reflecting light into camera.

BOOM. Crane-like device for suspending a microphone or a camera in mid-air.

BREAK. A point in a show where a commercial can be placed.

BREAKDOWN. Detailing costs for talent, props, etc. Also a script marked for camera shots by director of show or commercial.

BRIDGE. A short visual and/or audio sequence which connects portions of a drama or commercial.

BURN. When camera is held too long on bright image, it burns into the tube.

BUST SHOT. From chest up.

- C -

CATV - COMMUNITY ANTENNA TELEVISION. Television brought to the TV set by cable instead over the air. User pays a subscription fee.

CALL LETTERS. The actual name approved by the F.T.C., although most people refer to their local stations by the channel numbers.

CAMERA CARDS. Art cards of titles, credits, sponsor, etc.

CHANNEL. The frequency used by the station.

CHEAT. Varying positions to achieve a better TV picture. Two people may not actually face each other in studio, but may cheat somewhat to face more toward camera for better effect.

CHROMAKEY. A videotape technique where a person can be inserted over a different background. The person (or object) can be made smaller or larger.

CIRCULATION. The size of a TV audience. Can be broken down from TSA, total survey area, to smaller geographic parts. Also by age and sex.

CLIP. Previously shot film used in a new program.

CLOSED CIRCUIT. Television which does not go over the air, but is shown from camera to monitor.

CLOSE UP (CU). Showing details by moving camera in very close. Usually a portion of object is shown rather than the entire subject.

COAXIAL CABLE OR COAX. Cable used for the transmission of television signals.

COMMENTARY. Narration or voice over. Spoken description which complements the TV picture.

COMMERCIAL. Called a spot or announcement.

COMMERCIAL PROTECTION. Stations protect an advertiser by not having competitive commercials near each other in time.

COMPOSITE. The term used to signify both picture and sound on film.

CONFIRMATION. The okay by the advertiser.

CONTINUITY. The script.

CONTOUR CURTAIN. Curtain that can open via loops or scallops as usually seen in cinema houses.

CONTRAST. Ratio of light and dark portions of TV picture.

CONTROL ROOM. Area adjacent from studio where technicians operate.

COPY. Script.

C/P/M - COST PER THOUSAND. The cost to reach 1,000 households or people. Total audience in thousands divided by the cost equals C/P/M.

COUNT DOWN. The signal line in studio or on tape or film prior to the actual start. Film has an academy leader starting at ten seconds and counting down to zero.

COVERAGE. The area in households or prople reached by the station. The area is called the coverage area.

COW-CATCHER. An announcement at the start of a show advertising a product of the sponsor or another advertiser.

CRANE. Flexible type of camera mount that allows many different shot angles.

CRAWL. Lettered titles and credits that move up or from side to side on the TV screen. Usually at the close of a show.

CREDITS. People who produce the show. Usually shown on screen at the end of the show.

CUE. A signal by sight or sound to direct the start of show, music, narration or action.

CUT-BACK. To return to something previously shown.

CUT-IN. Often a local message inserted into the middle of a network show.

CU. Close up. A tight camera shot.

CUTS. Portions of script which are to be eliminated.

CYCLODRUM. Light inside a wooden drum with holes in its sides to give the appearance of flashing lights.

- D -

DMA - DESIGNATED MARKET AREA. An area of counties where the home market TV stations are dominant in total hours viewed. Stations from other markets might be viewed, but are not dominant. A term used by A. C. Nielsen. It is similar to ADI used by Arbitron.

DAMPEN. Soften hard-surfaced walls in studios with rugs or other materials to reduce noise in studio.

DAYPARTS. The parts of the day of telecasts such as morning, afternoon, evening, night, late night.

DEAD MIKE. An unconnected microphone.

DECIBLE. The unit of measurement of sound levels.

DEFINITION. The sharpness seen on a TV picture.

DELAYED BROADCAST. Broadcasting a program by a local station at a later time than the network program.

DEMOGRAPHICS. The age, sex, economic status and other characteristics of the television audience.

DEPTH OF FIELD. Where a performer may move, toward or away from camera, without going out of focus.

DIMMER. Equipment for changing the intensity of studio lights.

DINKY. A small spotlight used in TV.

DIRECTOR. The coordinator of production of a TV show or commercial.

DISHPAN. A circular transmitter.

DISSOLVE. An overlapping fade-out of one picture, fade-in of another.

DISTORTION. An abnormal change in sound or picture.

DOLLY. A movable fixture which holds a camera and can be wheeled about. Also, the movement of the camera.

DOWN-AND-UNDER. Instructions given to a musician or sound effects man to take down the volume and allow speech to be heard.

DOWNSTAGE. Area described as the front portion of the stage. In TV, toward the camera.

DRY RUN. Rehearsal prior to actual on-the-air performance.

DUB. A copy made from a video tape. Sometimes called dupe.

DUBBING. Recording one or more sound tracks on a single film or tape.

DUPE. A copy of an original film or video tape.

DUPLICATION. Reaching the same people with a specific TV schedule more than once.

- E -

ECU. Extreme close up. Coming in extremely close with the camera such as to show detail of the neckline in a dress.

ELECTRONIC MATTE. Mixing images from two cameras so the picture appears to be from one camera.

ELECTRONIC EDITING. Putting together video and audio or tape without manually cutting the tape.

ERASE. Wiping the video or audio.

ETV. Educational TV.

- F -

FACT SHEET. A checklist on things to do in the telecast. Also store buyer's information on merchandise to be advertised.

FADE. The visual fades to black or fades in from black.

FEED. A program fed from the network to station.

FEED BACK. The squeal or howl from accidentally closing the inbound and outbound ends of an electric circuit, or from improper mike hook-up.

FILL LIGHT. Lights used to eliminate unwanted dark areas.

FILM CHAIN. A film projector system.

FILM CLIP. Part of a film inserted into another film or show.

FILTERS. Lenses used to diffuse light and reduce glare.

FILTER MIKE. Microphone fixed to give special voice effects, such as a telephone voice sound.

FLAT LIGHT. Lighting of a TV set where the light is equal in all areas.

FLIP CARDS. Cards holding titles that may be flipped by stagehand while camera picks up image.

FLOOD LIGHT. A row of lights flooding the entire stage.

FLOOR MANAGER. Director's assistant, who is on stage and transmits information between control room and talent on stage.

FLUFF. Any mistake in action or words.

FM. Frequency modulation.

FREEZE. To stop the action.

FREQUENCY. The average number of times an unduplicated audience views a given TV schedule, programs or announcements or both, within a fixed time period. Example: A commercial aired a number of times over a four-week period will be seen an average of X amount of times by a certain total percent of the television audience. (See "Reach.")

FRINGE AREA. The edges of a geographic area reached by a station.

FRINGE TIME. The television dayparts of time before and after the more important dayparts -- such as late night.

- G -

GENERATION. With tape or film, the master is the first generation. A copy of it is the second generation. Copies of the latter are third generation.

GHOST. Double image.

GO-TO-BLACK. A slow visual fade to black.

GRAY SCALE. Differences from black to white.

GROSS AUDIENCE. The total homes or people for each television announcement.

GROSS RATING POINTS - G.R.P. The sum total of the rating points of households or people (or men or women, etc.) that are reached in a specific TV schedule.

- H -

HEADS AND TAILS. The beginning and end of any TV film or tape sequence.

HEADROOM. Height the camera can view above talent's head while still losing the mike and other extraneous items.

HIATUS. When an advertiser takes his schedule off the air for a period of time.

HIGH BAND. Tape recording of a high resolution versus low band.

HITCH-HIKE. An announcement of products which appears at the end of the show. Can be the sponsor of the show or some other advertiser.

HOMES. Homes owning one or more sets. Does not include sets in public or any place other than households.

HOMES USING TV - HUT. The percentage of homes viewing during a given time period.

HORIZONTAL SATURATION. Heavy schedule placed at the same time for several days.

HOT CAMERA. Warmed up camera ready for use. The red light is on.

HOUSEHOLDS. Same as homes.

- I -

INSERT. Picture inserted in another picture to strengthen an idea or show close-ups.

INTERCUT. Cutting from one camera angle to another to produce various shots of the same subject or scene.

IRIS. Transition between scenes beginning with a dot in mid-screen and iris-like opening to reveal new scene.

- K -

KEY LIGHT. The prime source of light.

KILL. To eliminate part of the action or film or tape.

KINESCOPE - KINE. Filming of a television show from a monitor. Kine is the abbreviation and is pronounced "kinny."

- L -

LS. Long shot.

LAPEL MIKE. Microphone attached to lapel.

LAVALIERE. Microphone around the neck.

LEADER. Portion of film that allows for threading and cuing.

LEVEL. The itensity of an audio or video signal.

LIMBO. Camera shot where subject has no frame or reference.

LIP SYNC. When the spoken word and lip action coincide with sound track.

LIVE. Transmission at the time of shooting.

LIVE COPY. Copy read by announcer at time aired.

LIVE FEED. A network feeding a program to its affiliates while it is being aired.

LOG. Breakdown of the entire day's broadcasting.

LOGO. Identifying symbol or signature.

- M -

MAKE GOOD. When an announcement does not run according to schedule, the station does a "make good" of a spot in equal or better time.

MASKING. Eliminating parts of the set.

MASTER. The original tape or film.

MASTER CONTROL. The controls in the control room of the station for transmission.

MATTE SHOT. A scene where a portion of picture is matted away electronically and another picture inserted in its place.

MICROWAVE. Frequency range. Used to carry TV signals through the air.

MIX. Combining and blending audio and video.

MOBILE UNIT. A studio on wheels. Signal can be microwaved back to the studio for broadcast.

MOCK-UP. Facsimiles of products. Can be same size or larger.

MODEL. A miniature of a scene or set made before the sets are finally constructed in full size.

MONTAGE. Combining several pictures together.

MOVIOLA. An instrument used to view film while editing.

MS. Medium shot in contrast to a close-up or a long shot.

MURAL. Photographic enlargement of a set to give the impression that the scene actually exists in the studio.

- N -

NETWORK. Stations which are affiliated with each other. Linked up by cable or microwave to allow simultaneous telecasting.

NETWORK FEED. Sending program material over the network cable to the affiliates. Broadcast can be aired at same time or delayed for later.

NIELSEN, THE A. C. NIELSEN COMPANY. A research organization which measures television audiences locally and nationally.

- O -

O & O STATIONS. Stations owned and operated by a national network.

OTO. One time only.

ON CAMERA. Talent or object is being televised.

ON THE AIR. Program in process of telecasting.

ONE SHOT TV. Picture of one person.

OPEN COLD. To open a show without a theme, or musical introduction, or even a rehearsal.

OPEN END. A filmed or video-taped program with commercial sections omitted to permit their being inserted at point of broadcast.

ORIGINATE. To issue a show from a particular location.

OPTICAL SOUND TRACK. A visible track that runs along the side of film. The sound and the picture are recorded at the same time.

- P -

PACKAGE. A complete program sold as a package.

PACKAGE PLAN. A combination of availabilities offered to an advertiser at a package price.

PAN. Moving camera around from side to side, or up and down.

PARTICIPATIONS. Commercials in a program not sponsored by the advertiser.

PIGGYBACK. An advertiser's commercial made up of two individual announcements for different brands or products, back to back.

PILOT. An example of a TV series to show advertisers how the series will look.

PLAYBACK. Replay of a recorded program.

POP IN. Technique where something pops on screen suddenly. Can also pop out in similar manner.

POWER. Amount of electricity a station operates on.

PRE-EMPTIBLE SPOTS. Commercials sold at reduced rates. The station can sell them to an advertiser willing to pay the full rate.

PRE-EMPTION. Replacing a scheduled program or commercial.

PROGRAM DIRECTOR - PD. In charge of the station's program operations.

PRIME TIME. 7:30 PM to 11:00 PM in Eastern, Mountain and Pacific Time Zones. 6:30 PM to 10:00 PM in the Central Time.

PRODUCER. The organizer of television shows or commercials and in charge of all financial matters.

PUBLIC DOMAIN. Belongs to the public. Can be used by anyone.

PUBLIC TV. Television supported by the public through funds and donations.

- R -

RATE CARD. Commercial price list.

RATE HOLDER. Time bought as part of a package plan to secure or hold a specific rate.

RATE PROTECTION. The length of time a certain rate is guaranteed at the agreed price.

RATING. The percentage of total TV homes in the coverage area measured, showing how many viewed a given time segment of television. Example: A 10 rating means that 10% of all TV homes in the area saw a particular TV program or announcement.

RATING AN AREA OF DOMINANT INFLUENCE - (ADI). This rating is based on those counties in which stations or originating market secure preponderance of viewing.

REACH. The net, unduplicated audience you "reach" within a time that you are measuring.

REAR PROJECTION - RP. A picture on screen projected from the rear allowing performers to work before the screen and appear as though they were on location.

RELEASE PRINT. Final videotape or film approved for on-air.

REMOTE. A telecast from outside the studio.

REPRESENTATIVE OR REP. A company which represents stations and sells national advertising for them.

RE-RUN. Repeating a show.

RESOLUTION. The amount of picture clarity.

ROUGH CUT. The first steps of film editing. Film segments are spliced together before adding special effects.

RUSHES. Film processed in a hurry for evaluation.

RUN-OF-STATION - ROS. Commercials get placed at station's discretion. Similar to run-of-paper with newspapers.

- S -

S.A.G. The union for actors performing on film.

SOF. Sound on film.

SOT. Sound on tape.

SANDWICH. Commercial with a set opening and closing. The middle can be changed.

SATURATION. Heavy use of commercials in a short period of time.

SCHEDULE. Time of day and dates an advertiser's commercials are on air.

SCRATCH PRINT. A film with a scratch down the middle so it cannot be used on air.

SECOND GENERATION. A copy of film or tape from the original master.

SETS IN USE. See Homes Using Television.

SHARE OF AUDIENCE. The percentage of the TV viewers tuned to a particular program compared to all people viewing TV at the same time.

SHOT. Picture produced by a TV camera.

SIGNATURE. The video and/or audio device used to identify a program.

SIGN OFF. When the station ends its broadcast day.

SIGN ON. When the station begins its broadcast day.

SIMULCAST. Televising a show and broadcasting on radio at the same time.

SLATE. A board which describes the program and those responsible for it.

SLIDE. A single frame film transparency.

SNOW. "Snow" on the TV screen. A weak signal.

SOLID STATE. Transistors instead of tubes.

SPLIT SCREEN. Divided screen to show two or more pictures.

SPONSOR. An advertiser who buys all or part of a program.

SPOT. Used as a synonym for announcement. Spot is also time bought market by market.

STAND BY. A cue that the TV program is about to go on the air.

STATION BREAK. Interval between shows, usually at 1/4, 1/2, 3/4 and on the hour. Used to identify the call letters of TV stations.

STEP IT UP. Increase volume of mikes or pace or tempo, action or music.

STILL. A stop-motion photograph.

STOP ACTION. Stopping a subject in motion.

STORYBOARD. Artwork showing the sequence of a TV commercial with its visual and audio changes.

SUPERIMPOSE. Overlapping two or more video sources to create one picture.

SYNC. Matching the voice to the sound.

- T -

TF. Til Forbid. Advertising schedule with no fixed expiration date. Runs until the advertiser terminates it.

TAG. Information added at end of a commercial.

TAKE-UP REEL. Reel on tape machine of film projector that receives used tape or film.

TELEPROMPTER. A device that rolls a script across a screen placed on the front of the camera. The performer refers to it.

TRAFFIC. Departments in stations that handle broadcasting of scheduled commercials.

TRANSCRIPTION. A recorded announcement.

TV SATELLITE. A station which picks up the signal of the mother station and transmits it to the satellite area.

- U -

UHF. Ultra High Frequency. Channels 14 to 83.

UNDUPLICATED AUDIENCE. Number of viewers reached at least once by a television schedule.

- V -

VHF. Very High Frequency. Channels 2 to 13.

VTR. Videotape recording.

VERTICAL SATURATION. Concentration of a heavy schedule within a day or days.

VIDEOTAPE. Magnetic tape used to record sound and picture at the same time.

VIDICON. A TV pick-up tube used in TV cameras.

VOICE OVER. Announcer's voice broadcast, but he is not seen.

ACKNOWLEDGMENTS

One chapter tells of the different people who are able to help you in television. The author, too, leaned on many to help him.

There are two famous companies in the field of television audience research. A. C. Nielsen and Arbitron (American Research Bureau). Considerable statistics are shown from Arbitron through their courtesy.

When it comes to learning more about what other advertisers are doing in television, TvB (Television Bureau of Advertising) is helpful to advertisers as a service from the industry. The "print converter" and other methods are shown through their courtesy.

Some storyboards are shown. Several were especially prepared by J. C. Penney for the Retail Advertising Conference. This is an annual meeting which takes place in Chicago. Other storyboard assistance is from Retail Broadcast Consultants. Thanks go to these companies.

Working with hundreds of stores and hundreds of people means that they enriched the author's television knowledge. Without their assistance, this book could never have been created. Instead of saluting them by name, he says "Thank you, all of you."

A special thanks should also go to Kenny Rothstein and Dick Kluga who helped in the realization of "his dream."